Book 1
Circuit Engineering
By Solis Tech

Book 2
Robotics
By Kenneth Fraser

Book 1
Circuit Engineering
By Solis Tech

The Beginner's Guide to Electronic Circuits, Semi-Conductors, Circuit Boards, and Basic Electronics

Circuit Engineering: The Beginner's Guide to Electronic Circuits, Semi-Conductors,
Circuit Boards and Basic Electronics

Circuit Engineering: The Beginner's Guide to Electronic Circuits, Semi-Conductors, Circuit Boards and Basic Electronics

Table of Contents

Introduction

I want to thank you and congratulate you for purchasing the book, **Circuit Engineering: The Beginner's Guide to Electronic Circuits, Semi-Conductors, Circuit Boards, and Basic Electronics**.

This book contains a beginner's course on circuit engineering. Here, the basics of electric and electronic circuits are discussed. You will grasp the definitions of circuits, semi-conductors, resistors, inductors, transformers, circuit boards, and electronics, in general. You'll even be introduced to electrical safety tips and a set of skills needed in electronics, as well as a short take on reverse engineering, hacking, microcontroller programming, and robotics.

Alongside, you can apply all that you'll be learning once you get started with the proposed circuit projects for beginner. You'll also be rewarded a peek at different career-advancement possibilities. While reading about the fundamentals and various theories in the subject is important, hands-on learning is equally important. This way, you can put your newly gathered knowledge to good use.

If you're uncertain whether or not you have what it takes to learn the ropes in circuit engineering, let this book help you decide. Chances are, you have the stamina for the field and for all you know, you can discover a new passion for circuits and electronic devices.

Thanks again for purchasing this book; I hope you enjoy it!

Chapter I – First Things First: An Introduction to Circuit Engineering

In 1882, there was a *circuit war*; it was between the notable electrical engineers and scientists, *Thomas Edison* (inventor of the DC system) and *Nikola Tesla* (inventor of the AC system).

While Thomas Edison stated that an efficient way of distributing power was via a *DC system*, Nikola Tesla argued that although DC systems are efficient, an *alternating current* is the *more practical* option. It started as a simple clash of ideas, but it eventually led to a major rift. Neither professional conceded; both of them insisted that their own systems were "better".

In the end, it was *Nikola Tesla* that took home the *glory*. Case in point? He was granted funds by an internationally recognized firm, Westinghouse. The majority of the power sources of New York City were based on the ideas of the Serbian engineer; at Niagara Falls in Canada, a power plant was built.

If you're interested in finding out more about the particular *circuit war, AC and DC systems*, and all critical discussions on circuits, taking a course about electronic circuits is the way to go.

I.A. - What Is a Circuit?

> Both an *electric circuit* and an *electronic circuit* refer to a complete pathway for electric current, which starts and ends at a single point; it is a passage that allows the electricity to enter at one place, then, let it pass through a series of stops, and finally, leave it to exit at the same place. The list of basic examples of a circuit includes a *light switch* (off and on) and *battery-operated lamps*.

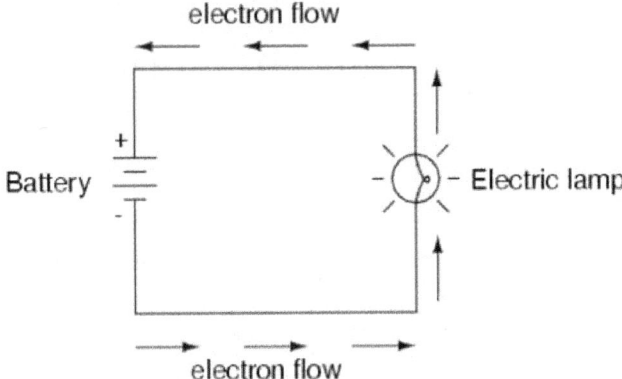

A circuit that follows a fundamental design

A circuit can function well - granted that its design is well-conceptualized. As much as possible, it is recommended that arriving at a simplistic product should be the goal; the simple and straightforward a design is, the better. With a fundamental concept, even if other (beginner-level) circuit engineers who will subject it to inspection will not have a difficult time in understanding its flow. Although there may be complex systems, the agenda is not intended to complicate the explanations.

Moreover, a circuit can be referred to as a space with a conductive path that grants electrons the opportunity to move freely. To create one with a brilliant design, a tip is to learn about the classifications of all circuits. You can use the knowledge to determine the appropriate kind of network, as well as the need for an external or internal source.

2 classifications of a circuit:

1. Linear or non-linear – a circuit that is based on either linear or non-linear networks; it is composed of independent and/or dependent sources and passive elements

2. Active or passive – a circuit that is based on either the absence (passive circuit) or the presence (active circuit) of a source; a source can be a power source or voltage source

I.B. - A Circuit & Its Types

Not all circuits are alike. In fact, one of the most common misconceptions involves an *electric circuit* and an *electronic circuit*; both are said to be one and the same, but they are not. While the former can carry *average to high voltage*, the latter has the tendency to have *low voltage load*.

Moreover, it is always important to be aware of the different circuit types, especially if you're about to make your own circuit; the kind of circuit that you create needs to have the ability to handle a preferred load.

Circuit types:

- Closed circuit – it is a circuit that is fully functional

- Open circuit – it is a circuit that can no longer function due to a damaged or missing component, or a loose connection

- Short circuit – it is a circuit that comes without a load

- Parallel circuit – it is a circuit that connects to other circuits; it is like the main power source or the primary circuit in a series of circuits

- Series circuit – it is a circuit that connects to other circuits; the same amount of electricity is distributed to each of its component circuits; the main power source or the primary circuit is unclear

I.C. - Conductors, Insulators & Semi-Conductors

Conductors, insulators, and *semi-conductors* give light to the fact that a circuit's electrical properties are dependent on the circuit type, as well as on their conduction bands (i.e. their allowed electric power). For instance, if a particular power source chooses to distribute a 9-volt electric power to a closed circuit, its electrical properties can be evaluated by using 2 details: (1) its characteristic as a closed circuit and (2) 9-volt electric power.

Moreover, conductors, insulators, and semi-conductors are integral concepts to the *conductivity* of an object. While conductors and semi-conductors are grouped to describe *charged carriers*, insulators are still considered as relative despite not containing any free charge.

Circuit Engineering: The Beginner's Guide to Electronic Circuits, Semi-Conductors, Circuit Boards and Basic Electronics

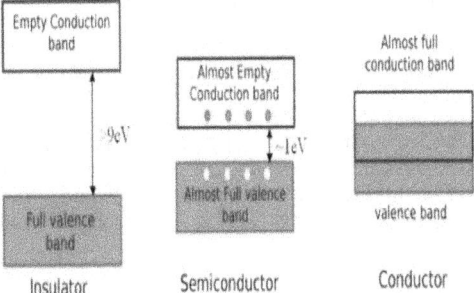

Insulators can be any "ion-less" object; the most common examples of semi-conductors are copper and aluminum & for conductors, gold and silver

Conduction bands:

- Conductor – it is a conduction band that is referred to as the almost full band

- Insulator – it is a conduction band that is referred to as an empty band

- Semi-conductor –it is a conduction band that is referred to as an almost empty band

I.D. - Breaking Down the Components of a Circuit

A circuit can be either *simple or complex,* and be both *simple and complex.* If the circuit in subject is a series circuit, with a group of 10 different circuits that are connected to it or if the said circuit is just a basic closed circuit with 5 different stops, it can be rather confusing to trace. However, if you dissect any circuit, you'll discover 3 *constant,* integral components.

Integral components:

1. Load – it is the representation of the power consumption, as well as the work that is accomplished within a system; without it, there's barely a point in having a circuit

2. Power source – it is where the electricity comes from

3. Pathway – it is the framework of a circuit; from the power source, it follows the load through each of the network, and finally returns to

and exits the power source; it is also referred to as the conductive pathway

I.E. - The Roles of Current, Resistance & Voltage

Current, resistance, and voltage are the 3 representations of the important components of a circuit's system. They can explain how electricity enters, then, moves from 1 point to another, and finally exits. Whether the path of electricity is rather simple, these representations remain constant. Apart from describing the electric flow, they can serve as indications of faults (in instances when a circuit fails to work).

3 representations:

- Current – it is the representation of the electric flow; particularly, its focus is on the flow of electrons

- Resistance – it is the representation of the nature of an electric flow as it moves around the circuit

- Voltage – it is the representation of the electric force or pressure; in general, the supply comes from an electric outlet or a battery

I.F. - AC/DC Systems: Which System Is in?

AC and DC systems (or alternating current and direct current systems) are often associated to each other. When the *AC system* is mentioned, so is the *DC system.* Conversely, when it's the DC system's turn to be in the spotlight, it won't be long until the AC system is mentioned. This is because these systems are opposite of one another; to get a better understanding of one of them, it's recommended to be familiar with the other, as well.

Moreover, *AC and DC systems* are types of a circuit's current flow. In an AC system, the current flow changes its direction occasionally. Meanwhile, in a DC system, the current flow follows a single direction.

It can be deduced, therefore, that an AC system grants a circuit freedom to let the current flow in several directions. While this can be an advantage, this doesn't permit the continuous flow that a DC system can entitle.

So, should you use an *AC system* or a *DC system*? The decision as to which current system is dependent on the more practical design to follow; take into account the aim of having your own circuit. If you prefer something grand and you intend to power something large, the AC system can step in. On the other hand, if you're good with a basic setup, you can use the DC system's concept as basis.

I.G. – What Is a Transformer?

Circuit Engineering: The Beginner's Guide to Electronic Circuits, Semi-Conductors, Circuit Boards and Basic Electronics

A *transformer* is a device that serves as a portal for energy transfer within the points in a circuit or from circuit 1 to circuit 2. In most cases, it is used for increasing and decreasing the voltages in a system.

When the first transformer was built in the mid 1880s, circuit engineers discovered that a transformer significantly improves the electric flow in a circuit, and consequently, results to a more powerful circuit. The discovery made way for various transformer designs, as well as various transformer sizes.

A primary principle of a transformer is its need for extremely *high magnetic permeability*. It follows that a circuit that is capable of attracting power is, of course, more inclined to have electric current transferred to it; and, conversely, a circuit with *low magnetic permeability* is less likely to extract power from another circuit.

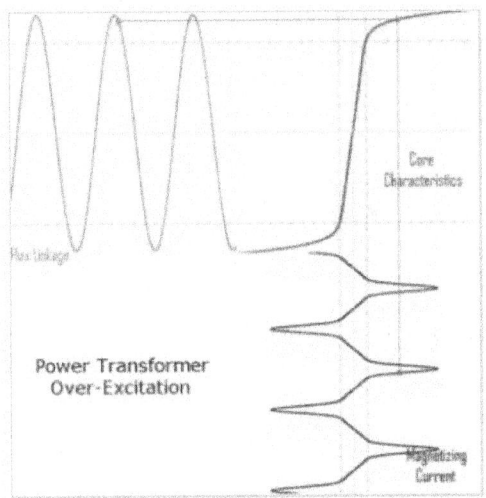

Magnetic permeability is defined as the ability of a circuit to hold and support an internal magnetic field

Chapter II – The Anatomy of a Circuit

Let's say that a lamp in your room has been around for a while, and let's say it chose to give up on you the other night; although it's an old lamp, you still believe it can *work fine*. You suspect that the problem is in the switching board. Instead of just letting it be and since you're interested in how it works, you choose to study its components.

As you open its internal system, you see that connected to some sort of panel are 2 wires; one wire is red, the other is black. As far as your knowledge in circuits can tell you, the 2 wires in your hands are: (1) a live wire or the wire that is connected to a switch, and (2) a neutral wire or the wire that carries the load.

As the universal rule in circuit engineering goes, *red* is the color that indicates a live wire, *black*, on the other hand, is a neutral wire. For a chance to know why your lamp gave up on you other than old age, checking the red wire would be a good start.

II.A. – Individually Speaking: Parts of a Circuit

A circuit can *work smoothly* if its individual parts are contributing to the workflow as expected. Remember that it follows a series; if one point in that series is not in condition. So, if you're wondering why a system is functional, check each one of its components. To use the old adage (and modify it a bit), the circuit, as a whole, is just as good as its individual parts.

Individual parts:

- Capacitor – it is in charge of the stability in the power source in a circuit

- Diode – it is in charge of supplying the light in a circuit

- Inductor – it is a coil of wiring system

- Resistor – it is in charge of power consumption

- Transistor – it is in charge of the electric signal control

II.B. – Circuit Categories: Which Category Do You Belong to?

Circuit categories describe the *voltage levels*, as well as their *electric flow over time*; they refer to how much and how powerful the current is as it enters and passes through each point in a circuit. Since they vary, it's listed as an effective way of understanding circuit systems of sorts if they are categorized.

Circuit categories:

- Analog circuits – these are the categories of circuits that use the concepts of parallel circuits and series circuits as basis. Among their fundamental parts are capacitors, diodes, resistors, and wires.

  In diagrams, analog circuits are easy to recognize. Usually, when a model of the particular circuit is drawn, a simple illustration is presented since these circuits do not follow a complex system. In most cases, when illustrated, the parts (e.g. capacitors, diodes, wires, etc.) are represented by lines.

- Digital circuits – these are the circuits that rely heavily on Boolean algebra (i.e. values are either true or false, or as denoted 1 or 0); therefore, these circuits are often dependent on transistors that can create closed logic. Compared to analog circuits, these follow a state-of-the-art design.

  Furthermore, digital circuits are designed to create either numerical or logical values for the representation of electricity that flows within their system. Since these are not only focused on the mere ability to take in electric current, but rather, on the individual properties of all of their parts, as well, these circuits come with advanced functions; they can provide memory and accomplish arbitrary computations.

- Analog-digital – these circuits are sometimes called hybrid circuits or mixed signal circuits since both system designs of analog and digital circuits are given light. Although their concept can be quite complex, these circuits can deliver a more thorough result; the procedures are combined, which allows collaborative effort from different parts. One example is a telephone receiver; first, it works based on analog circuitry to create and stabilize signals, then, based on digital circuitry, these signals are converted into digital units, and finally, are subjected to interpretation.

II.C. – Where Does Inductance Enter the Picture?

In circuit analysis, the term *inductance*, introduced in 1886 by *Oliver Heaviside*, refers to the property of a circuit's electricity-producing component to change in amount. Apart from the support of a circuit's aspect to vary, it points out the need for a filter and energy storage systems to be provided.

As it follows, the component in a circuit that enables inductance is called an inductor. Usually, these parts are made out of wire. But, while some circuits contain inductors as integral parts, others remain functional without the need to alter the electric flow.

Inductance can be either *mutual inductance* or *self-inductance*. The former refers to a change in electric current from one inductor to another inductor; it explains the primary operations of a transformer. Meanwhile, the latter refers to the *stable inductance* within a system.

Moreover, inductance is represented by the symbol L, which is meant to giver to the scientist, *Heinrich Lenz*. It also measured in units of *henry* after the American scientist, *Joseph Henry*; it follows that although it was Oliver Heaviside who introduced the term, the man behind the development is Joseph Henry.

Mutual inductance describes the occurrence in a circuit when there is change that can be traced to an inductor; particularly, it refers to the alteration due to an inductor's preference of a nearby inductor. It is essential to learn about the relationship of 2 *inductors* since it is the basis of the operations of a transformer. Additionally, a limit has to be maintained in order to keep potential energy transfers regulated; the failure to incorrectly calculate mutual inductance can result to *unwanted inductance coupling*, as well as a *power overload*.

It can, therefore, be deduced that mutual inductance (as represented by the symbol M) is the measurement of the coupling that involves 2 inductors; the 2 inductors are, then, given particular importance (in terms of coil turns), along with each inductor's ability to admit current flow or *permeance*. The formula for calculating mutual inductance is as follows:

**representations: M_{AB} = mutual inductance in circuit A and circuit B

N_B = inductance in circuit B

N_A = inductance in circuit A

P_{AB} = permeability in circuit A and circuit B

$M_{AB} = (N_B)(N_A)(P_{AB})$

When explained, the formula highlights that the mutual inductance between inductor A and inductor B (or M_{AB}) is equal to the product of 3 elements: [1] the coils of inductor B (or N_B), [2] the coils of inductor A (or N_A), and [3] the permeance of inductor A and inductor B (or P_{AB}).

On the other hand, self-inductance refers to voltage induction of a current-carrying system in respect to the changing current in the circuit. It points out that eventually, there will be another current that will flow along with the primary current. Due to the amount of force within the magnetic field, voltage is induced; particularly, voltage is *self-induced*.

II.D. - What Makes an Integrated Circuit?

An *integrated circuit* (or IC) is alternatively called a microchip or a chip, due to its size. It works depending on a particular signal level. One example is the integrated circuit that enables a computer to perform a multitude of tasks; instead of loading a computer's structure with a large circuit, it comes to the rescue.

In most cases, an *integrated circuit* operates at little defined states. Compared to the normal circuit whose operations are distributed over continuous amplitudes, it can function within a small network; the normal circuit may sometimes fail to work with only minor amplitude ranges.

Basically, an *integrated circuit* is no different from any other circuit; its power can astound you, yes, but, if it comes down to describing how it is, it's simply a circuit that has been reduced so it can fit inside a chip.

Chapter III – Resistance Isn't Futile

Without a material that can act as the opposing force, a circuit can *function*, but it may not function *as desired*. When an electric supply can perform its function by distributing electricity to the opening of a circuit, the electric current will keep on flowing; its flow can be uncontrollable, which can destroy a system's integrity. Usually, without the opposition, a circuit ends up taking too much load.

The term for this opposing material is *resistance*; it goes hand in hand with the term conductance. And, as mentioned in the first chapter, it is the representation of the current flow in a circuit.

III.A. – What Is Resistance?

Resistance is the measurement of an opposing electric current; it can be expressed in ohms. It generates an amount of friction that is relative to the necessary amount of electricity that a particular circuit can handle.

In a way, *resistance* is responsible for the *smooth flow of electricity* in a circuit. Although others would counter the argument by saying that rather than support the effortless flow of electricity in a system, it slows it down.

However, it is *resistance* that allows *balance of electricity* in a circuit. Take for example the case of a circuit that can only handle a total of 15V. If a circuit takes in 20V at a resistance of 5 ohms, the number is diminished to 15V, which indicates a functional circuit. Conversely, if, in the same situation, there is no *resistance* of 5 ohms, a circuit may not be as functional as desired; its system ends up carrying 20V, which implies that it is overloaded.

III.B. – Resistive Circuit 101

A *resistive circuit* is a kind of circuit that consists of nothing but a series of resistors to complete the combo of electric current and voltage source. If viewed in a chart, it is noticeable that the power waveform is always positive; it is suggestive the power in a circuit is always dissipated, and is never returned to the original source.

It is important to note that the frequency of the power in a circuit should not be equal to the frequency of the electric current and voltage. If possible, the frequency of the power should be twice as high as that of the electric current and voltage. This unequal frequency distribution grants constant change within a system.

Since it is made up of resistors and does not include transistors and capacitors, a resistive circuit is rather easier to analyze. Understanding the electric flow within the circuit (whether in an AC or DC system) requires a straightforward technique. Therefore, determining the flow of the current in a resistive circuit is simple; by adhering to the formula, calculating the figure is easy.

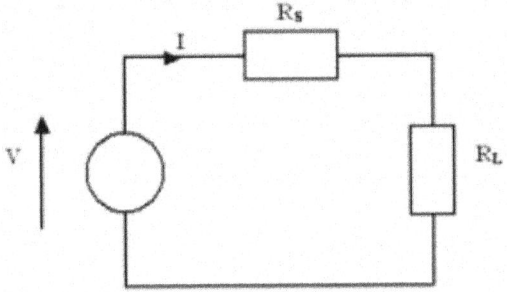

In a resistive circuit, voltage can easily be monitored

**representations: I = total current

RS = Resistance Source

RL = Resistance Load

I = voltage ÷ (RS +RL)

III.C. - War between the Types of Resistance

Resistance is classified according to the type of resistivity that it can contain, along with the amount of resistance that a circuit can carry. This allows the opportunity for an opposing force to be valued, regardless of its resistivity. As the professional electrical engineers can attest, not every circuit component that produces resistance satisfies the rules, particularly, *Ohm's Law.*

2 types of resistance:

- Differential resistance – it is the resistance derivative of voltage in light with the electric current; also referred to as *incremental resistance, small signal resistance, or dynamic resistance*, its concept is responsible for oscillators and amplifiers

17

- Static resistance – it is the resistance that corresponds to the typical definition of resistance; it is also called *chordal resistance* or *DC resistance*

III.D. - Resistance vs. Conductance

The average circuit comes with both *resistance* and *conductance*, which gives balance to the electric flow in a circuit's system. While the former refers to the opposition, conductance describes the amount of current that is converted into power that revolves around different points.

Conductance also covers the ability of a circuit's components to conduct electricity. And, to bring light to its counterpart's ability to oppose the flow, it dwells on the subject of the convenience of electricity to pass through a series of points in a circuit.

With both the resistance and the conductance in the system, a circuit can function as desired.

III.E. – The Need for Calculations (Four Ways)

For electric current to flow smoothly within a system, a level of resistance has to be present. And, since not all circuits come with a similar design, their resistance levels vary. To calculate a particular circuit's resistance, first, you need determine its type, and its provided values, as well.

Four ways:

#1 – Resistance calculation for a series circuit

The formula:

Resistance = $_1R +\ _2R +\ _3R +\ _4R$

#2 – Resistance calculation according to voltage & power

**representations: total voltage; PT = total power

The formula:

Resistance = $VT^2 \div PT^2$

3 – Resistance calculation according to voltage & current

**representations: VT = total voltage; IT = total current

The formula:

Resistance = VT ÷ IT

4 – Resistance calculation according to power & current

**representations: PT = total power; IT = total current

The formula:

Resistance = PT ÷ IT

III.F. – What about Sheet Resistance?

Sheet resistance refers to the measurement of the resistance in a thin sheet in a circuit's components. It can be used to describe the resistibility of different circuits and can point out the specific difference in circuits that vary in size. Especially in the case of a commercial product, the topic is covered for the assurance of quality.

You can look at *sheet resistance* as a special kind of resistance since it generates a more specific value. Usually, the average resistance in a circuit is expressed in *ohms*; *sheet resistance* is expressed in *Ohms per square*.

In most cases, *sheet resistance* is used for the analysis of circuits with uniform conductivity or semi-conductivity. Typical applications are extended to quality assurance for a commercial circuit.

III.G. – The Role of Impedance & Admittance

Like resistance, *impedance* can be described as the opposition in a circuit; unlike resistance, however, it refers to the opposing force of a circuit after the application of voltage. It is only relevant to AC systems or circuits where direct current isn't the supplied.

It was in 1893 when the concept of *impedance* was initially introduced by the Irish engineer, Arthur Kenelly. Back then, it was denoted by *Z* and is defined as a complex number.

When it comes to quantitative terms, *impedance* refers to the ratio of voltage to the electric current in a circuit. Its introduction is important for beginners especially if they're scratching their heads as to why there's an opposing force besides resistance.

Impedance, like resistance, comes with values. In a single open circuit, its value is presented in *ohms*. In the event of a series circuit or a parallel circuit, its value can be calculated by simply adding all the defined values in each unit.

The formula:

**representations: TZ = total impedance

Z_1 = impedance in component 1

Z_2 = impedance in component 2

Z_3 = impedance in component 3

Z_{10} = impedance in final component of a circuit

$$TZ = Z_1 + Z_2 + Z_3 \dots Z_n$$

Meanwhile, *admittance* is a relative concept in circuit engineering. It addresses the issue that alongside the difference in the magnitude of the electric current and voltage that are flowing within a circuit, the difference in phases needs to be given light, too. This way, the maximum load within a system can be calculated accordingly.

The formula:

**representations: TY = total admittance

Y_1 = admittance in component 1

Y_2 = admittance in component 2

Y_3 = admittance in component 3

Y_{10} = admittance in final component of a circuit

$$TY = Y_1 + Y_2 + Y_3 \dots Y_n$$

Chapter IV – It's Time to Measure the Electric Flow in a Circuit

In the previous chapter, the formulas for the calculation of a circuit's resistance levels were shared. However, the formulas for the *calculation of the entire load* in a circuit have yet to be discussed.

This is due to the significance of using the appropriate measurement units. In a few cases, especially those who are still on the initial phase of learning circuits? They aren't quite careful with their selected units. As it follows, it's not only necessary to calculate a circuit's electric flow; it's also necessary to calculate a circuit's flow correctly.

IV.A. – Standard Units

Among the several reasons to use *standard units of measurement* are for indications of exact measurements and for indications of the preferred measurements in a system. These units bring uniformity.

In circuits, the usual standard units that you encounter are *V, W, I,* and *P*. Although there are more, those who wish to explain a system's electric flow rely on these measurements; rather than introduce a bunch, which may only make matters more confusing, some are preferred. Moreover, without such units, understanding others' discussions of circuits is nearly impossible.

Standard units:

- Conductance – its measuring unit is Siemen with G as symbol

- Current – its measuring unit is ampere with I or i as symbol

- Frequency – its measuring unit is Hertz with Hz as symbol

- Inductance – its measuring unit is Henry with H or L as symbol

- Power – its measuring is watts with W as symbol

- Resistance – its measuring unit is ohm with R as symbol

- Voltage – its measuring unit is volt with V or E as symbol

IV.B. – Commonly Used Alternatives

Other than the *standard units of measurement*, other units are given light since these can enable clearer expression of the electric flow in a circuit. Especially if the circuit in subject contains a rather complex system, it can be difficult to arrive at a definite solution.

Other units:

- Angular frequency – it is a unit of measurement used in an AC circuit; it is a rotational unit that describes the relationship of at least 2 electric forms in a circuit

- Decibel – it is a unit of measurement that represent the gain in either current, power, or voltage; since it is only a tenth of the original unit, *Bel*, it is primarily reserved for denoting extremely small amounts

- Time constant – it is a unit of measurement that describes the output of a circuit's minimum or maximum output value; in a way, it refers to the measurement of time reaction

- Watt-hour – it is the unit of measurement that describes the electrical energy consumption over a period of time

IV.C. – Units of Force

Since *force* in a circuit is an important concept, it is advised that the particular unit of measurement is presented correctly. Even in physics, it is reiterated that it should be labeled according to its right category.

Atomic and electrostatic units of force:

- Hartrees

- Newtons

- Tesla

- Coulombs

- Meters

Chapter V – Power Transfer at Max

There will be instances of a *power transfer* in a circuit. For the electricity to continue its smooth flow, its original energy source will be replaced with an internal energy source. With such a change (especially in the case of a series circuit where power needs to flow continuously), the explanation for its system's pattern becomes a notch challenging.

For *beginners*, a great way of understanding *power transfer* in a circuit is to understand maximum power transfer, along with the concepts that dwell on the topic. As it follows, by gaining clarity on how much was the original power, as well as how much power a particular circuit can handle, you can see whether a power transfer is necessary or will only cause its load to be compromised.

V.A. – Maximum Power Transfer

Maximum power transfer, a concept that was introduced by Moritz von Jacobi sometime in the 1840s, draws light on the idea that for maximum external power to be obtained, an internal resistance needs to be in place. However, the transfer can only be flawless if the original resistance is equal to the potential power that an internal resistor can produce.

Consequently, maximum power transfer yields results that point out *power transfer*, and not *efficiency*; while improved efficiency can be a byproduct, it is not the chief purpose of maximum power transfer. It implies that although higher percentage of power is transferrable, it does not affect the magnitude of the power load (i.e. the extent that it can affect a circuit). In the event that the internal resistance is modified to accommodate a value higher than the value of the original resistance, improved efficiency can be achieved.

Moreover, the concept of maximum power transfer was initially misunderstood; a subject of many arguments was a circuit's reduced efficiency with the occurrence of transfer. Some insisted that due to the potential power that is lost during an exchange, a circuit may fail to reach 100% efficiency. As emphasis of this group's angle, take for example the case of a motor whose power is transferred from a battery; power in this situation may not be maximized, and it will only be realized over time when battery power has been fully consumed.

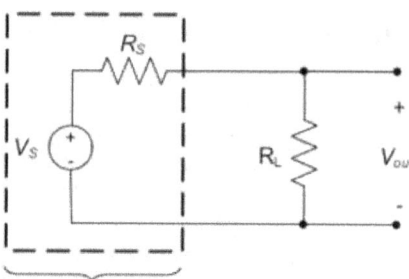

The maximum power theory states that the task of transferring power consumes power, too

It was *Thomas Edison*, as well as his fellow scientists, *Francis Robbins Upton*, who contested that maximum power transfer and efficiency are only *relative*; the 2 concepts are not one and the same. In fact, there is a discussion about *maximum power efficiency*, too.

In the exchange, you will find that *resistance* plays an important role. By giving light to the former argument, you can calculate a circuit's capability of a maximum power transfer, in relation to maximum power efficiency with the following formula:

MPT = RL ÷ (RL + RS)

A circuit's *MPT* (or maximum power transfer) can be determined with basic arithmetic skills. First, divide the *RL* (or Resistance Load) by the sum of the *RL* and *RS* (or Resistance Source).

V.B. – Thevenin's Theory

Thevenin's Theory, conceptualized by Hermann von Helmholtz and Leon Charles Thevenin, discusses that if a circuit follows a linear network, any point can be replaced given that it remains to carry a source for current, resistance, and voltage. Behind it, the idea is to supply an equivalent.

Originally, *Thevenin's Theory* can only be applied to circuits that operate with a DC system; since a DC system is rather simple, replacing its components with an equivalent is possible. Eventually, however, its capability to handle a load in a non-linear system was discovered; it can offer solutions for an AC system.

Moreover, the *Thevenin Theory* puts emphasis on that the average circuit can only be considered to have a linear according to a limited range; it can only be replaced by the components of with values among the range.

The Thevenin Theory follows that power dissipation can yield unique values, and can also yield identical values. However, the results can only be accomplished with the power supplied by an external resistor.

V.C. – The Star Delta Transformation

The *Star Delta Transformation* dwells on the idea that a circuit's system can change from one phase to another. For instance, if a circuit's power source is altered, its ability of carrying power from a point to the next is altered, too. Especially if there are 3 branches in a circuit's system, the power that circulates is known to form a closed loop.

The Star Delta Transformation refers to 2 kinds of circuit transformations. The first circuit transformation is a star transformation, which can be described by a "Y" formation; the second circuit transformation is a delta transformation, which can be described by a triangular pattern.

Moreover, the Star-Delta Transformation describes a 3-phase network of circuits, which can explain power transfer between these 3 networks. It enables the conversion of impedances that are connected to each other. With the theory as basis, alongside getting a clear scope for power transfer analysis, solving various concerns can be accomplished, too; the concept is applicable to different types of circuits including *series circuits, bridge-type networks, resistive circuits,* and *parallel circuits.*

The Star-Delta Transformation can be converted to the Delta-Star Transformation. From the star or Y-formation, the circuit creates a triangular network as the transition is achieved.

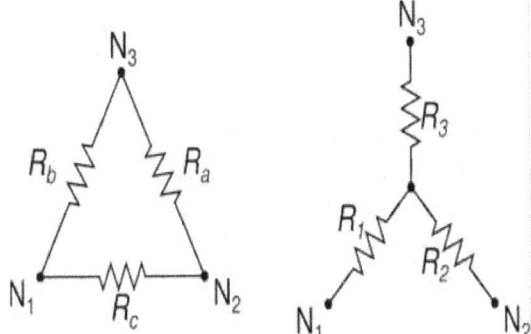

The Star-Delta Transformation or the Delta-Star Transformation is also called the Y-Δ Transformation or the Δ-Y Transformation

For the transition of the Star-Delta Transformation into the Delta-Star Transformation, a formula should be followed; this is meant to ensure that the transformation, along with the calculations for the total resistance in *all 3 circuits*, is successful. Initially, the goal is to compare the amount of power in an individual network. Once the power in network 1 has been acknowledged, proceed to identifying the weight that one network holds in the entire formation; one way of determining this is to disconnect that entire network and observe the operations of a circuit.

The formula:

**representations: $_\Delta R$ = total resistance of the transformation

$_1N$ = resistance in node 1

$_2N$ = resistance in node 2

$_AR$ = resistance in circuit A

$_BR$ = resistance in circuit B

$_CR$ = resistance in circuit C

$_\Delta R \, (_1N \, _2N) = _CR \, || \, (_AR + _BR)$

The simple version of the formula:

**representation: TR = total resistance

$_AR$ = resistance in circuit A

$_BR$ = resistance in circuit B

$_CR$ = resistance in circuit C

$_TR = _AR + _BR + _CR$

V.D. – Extra Element Theory

A circuit analysis technique that can be used for the simplification of a complicated problem is *the Extra Element Theory*; it was proposed by R.D. Middlebrook. The idea behind it is to take a complex matter, then divide it into small portions; each of the small portions will be addressed.

It follows that every circuit has a *transfer function* and *driving point*; the process of analyzing a circuit, therefore, can become easier if the aforementioned elements are first identified.

In the Extra Element Theory, unlike in other circuitry theorems, an element such as a capacitor or resistor can be temporarily removed so the transfer function or driving point can be determined. Since there are circuit components that can complicate an equation (regardless of how integral they are to a circuit), it is practical to set them aside for a while; although they may be of value to a circuit as a whole, it was proven that they don't affect calculations. Once the initial goal is achieved, the elements can be returned.

Impedance is a familiar term in discussions of the Extra Element Theory; it can be analyzed with the employment of the theory. In certain cases, its input can be determined in network granted that an *extra element* joins in.

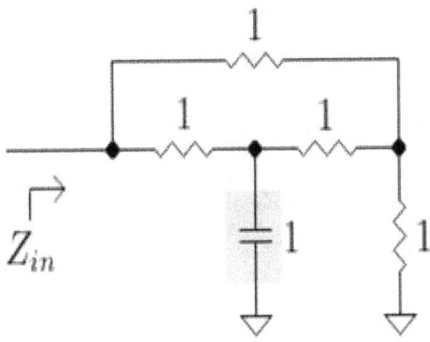

The Extra Element Theory (in relation to input impedance) proposes the addition of an "extra" element that is equivalent in value to the other elements

The formula for finding the impedance is:

**representations: Z = impedance

s = source

Z = 1 ÷ s

V.E. – Simplification of the Source

Simplification of the Source, sometimes referred to as Source Transformation, is the process of converting electric current into voltage,

27

or voltage into electric current. It is a common technique used by many circuit engineers for explaining their circuit's system in simple terms.

The process of Simplification of the Source usually begins with an existing resistance source in a circuit; it is then replaced with new electric current source with a similar level of resistance. Since it is *bilateral procedure*, one can be derived to yield results from another. It makes way for the adjustment of voltage as it gradually becomes the equivalent of a particular circuit's resistance.

Moreover, *Simplification of the Source* may begin with an existing resistance, but is *not limited to the accommodation of resistive circuits*. It means that the process can be performed on circuits that involve inductors and capacitors.

V.F. – Where Does the Rosenstark Method Fit in?

The *Rosenstark Method*, sometimes called *Asymptotic Gain Model*, is yet another important subject where power transfer is concerned. In light of the return ratio, it serves as the representation of *negative gain* from feedback amplifiers. As it provides an intuitive form of circuit analysis, it introduces a new batch of elements such as the *return ratio* and *asymptotic gains*.

The Rosenstark formula:

**representations: G_o = 0 asymptotic gain

G_∞ = infinite asymptotic gain

T = return ratio

Rosenstark Method = $\{G_o + [T \div (T + 1)]\} + \{G_{\infty} + [T \div (T + 1)]\}$

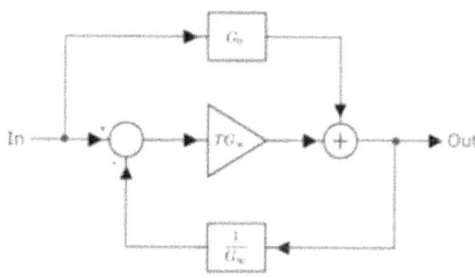

Circuit Engineering: The Beginner's Guide to Electronic Circuits, Semi-Conductors, Circuit Boards and Basic Electronics

The Rosenstark Method serves as the basic representation of power transfer in a circuit

Best features of the Rosenstark Method:

- Assumes that the direct transmission is small and can be equated to the asymptotic gain

- Characterizes bilateral properties and feedback amplifiers

- Identifies (via thorough inspection) the passive circuit elements

Steps:

1. Choose a source in a circuit's system; preferably, opt for a dependent source.

2. Determine the source's return ratio.

3. Identify the G_∞.

4. Identify the G_0.

5. Use the Rosenstark formula and substitute values.

Chapter VI – Laws, Laws & More Laws

Come to think of it, *circuit analysis* is one of the broad branches of electrical engineering. It covers the basics of circuits and power consumption; it also covers the extensive aspects of the topic such as the operations of an entire electrical network.

Without laws regarding how electric current is distributed within a circuit's components, there is a risk of unclear discussions of *fundamental and in-depth analysis*; without them, a guide that can break down lengthy components is absent. For beginners, especially, they are important since they grant the chance to understand the functions of a circuit and even each one of its parts. It stays true to the idea that for the resolution of a big problem, looking at it like little pieces is an opportunity to conqueror its difficulties.

VI.A. – Putting Kirchhoff's Laws into Motion

Kirchhoff's Laws are composed of *2 equations* that dwell on the conservation of charge and energy. They were initially introduced in 1845 by Gustav Kirchhoff as a means for the determination of a circuit's power consumption, as well as its parameters.

As several discussions go, Kirchhoff's Laws may not be too important since it was derived from the work of Scottish physicist, *James Maxwell*; since it was derived from another's work, it is said that a circuit engineer should rather use the original work as basis. However, it is a set of laws that primarily focuses on the operations of a closed circuit; the previous work from which it was taken doesn't put emphasis on a closed circuit, and is rather descriptive of the generic circuit.

Moreover, Kirchoff's Laws are conditional. It may be useful in describing charge and energy conservation for different circuit elements, but it can only yield an *approximation*; it also requires certain factors such as changing electric currents, voltage, and resistance.

Kirchhoff's Law is actually a general term. To be specific, it addresses 2 subjects: current and voltage. It acknowledges all circuits, regardless of complexities.

To solidify its argument in the conservation of charge and energy, one of Kirchhoff's Laws, *KCL* (or Kirchhoff's Current Law), states that the electric current in an interconnected network is relative. And, to emphasize his point, Gustav Kirchhoff contests that the algebraic sum of the electric current in a joint network is *0*.

Another law of Kirchhoff that complements the KCL is *KVL* (or Kirchhoff's Voltage Law). Its basis is on the general law on energy conservation that defines voltage as the energy per unit of charge. Just like in the previous law, Gustav Kirchhoff says that in a closed network, the algebraic sum of the electric voltage in a joint network is *0*.

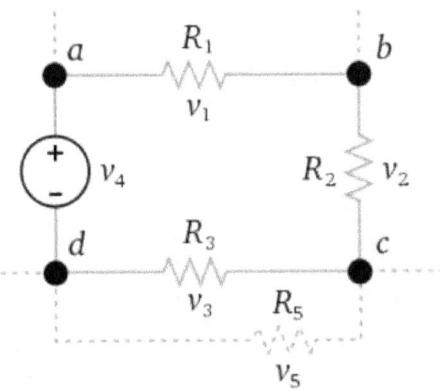

According to KVL, adding all the voltages within the circuit yields 0

VI.B. – Does Michael Faraday Know What He's Getting Into?

The *Law of Induction* by *Michael Faraday* is a fundamental law that revolves around the field of a circuit and the possible elements that are subject to eventual interaction. It carries both qualitative and quantitative aspects, and proposes that only with the presence of an infinite source of inductors can a circuit retain its inductive capability.

Unfortunately, not all circuit engineers and scientists are one with *Michael Faraday* and his *Law of Induction*. It is argued that although it holds some matters true, especially where loops of wire are concerned, it can yield the wrong results if used extensively; it can only handle a certain field and is usually arbitrarily small. In fact, counterexamples were previously presented.

A counterexample of Faraday's Law of Induction is the case involving an *electric disc generator*. Due to its own magnetic field, it can rotate circularly at a specified angular rate; it can complete a rotation and in the process, induct electricity by distributing it to other areas (within the circuit). It can be deduced that although the circuit's shape has remained constant over a period of time and although it can induct electricity, its distributive method made it lose inductive capabilities.

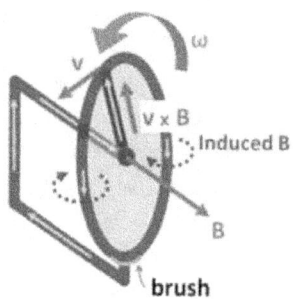

An electric disc generator induces electricity until cycle completion in the object's lower brush.

Faraday's Law of Induction may have been noted for certain flaws, but it is a law that made the conception of other important laws (not all of them are in the field of circuit engineering) possible. An important law with basis on Michael Faraday's idea is Albert Einstein's *Special Relativity*.

VI.C. - Ohm's Law & Conductivity

Ohm's Law, named after the German scientist, Georg Simon Ohm, is a powerful law that describes the resistance and electric current in a circuit, as well as their potential difference. Particularly, it sheds light on 2 points, then, subjects them to further analysis. Consequently, the relationship between Point A and Point B will indicate each of their properties as individual components.

In the verge of learning its concepts, it will dawn on you that *Ohm's Law* is an *empirical matter*. After observations with different ranges of length scales, it was proposed that it would fail where small wires are concerned; however, this proved to be a mere assumption since it was discovered that it could function regardless of a wire's size.

Since it is a basic equation in circuit engineering, *Ohm's Law* can also be applied for the determination of metal conductivity; it can be relied on when it comes to understanding the electric flow and conductive aspect in a circuit's component. Various materials that make use of Georg Ohm's principles are referred to as *ohmic*.

Georg Ohm set the parameters for his law to include only common terms in circuitry

In the physics of electricity, Ohm's Law is important where quantitative measurements in a circuit are concerned. Initially, when it was presented to German scientists, it was only mocked and rejected due to the mere fact that it went against the basic understanding of electric flow; it was dismissed as "a web of fancies" and worse, Georg Ohm was dubbed as a "fancy but invaluable" science professor. It was only until 1840 when it earned recognition and is now widely used today.

How to calculate for the elements in Ohm's Law:

Formula # 1:

voltage = current* resistance

Formula # 2:

current = voltage ÷ resistance

Formula # 3:

resistance = voltage ÷ current

VI.D. – The Argument of Norton's Theorem

Edward Lawry Norton, as he implies in *Norton's Theorem*, proposed that in circuit analysis, any linear network, granted that it has sources of voltage, electric current, and resistance, has an equivalent in dual network. A certain system may be unique, but values that can make a similar model can be obtained.

To find a circuit's equivalent using Norton's concepts, certain values need to be identified: voltage, electric current, and resistance. Once there are clear sources, you can begin formulating the necessary equations. However, when the sources are dependent on each other, another (and it has to be from a generic source) electric current needs to enter the picture.

Moreover, Norton's Theorem gives light to the fact that the values for the *equivalent current* and *equivalent voltage* are values that can be identified at the first 2 terminals of a circuit. It is important to note, however, that such a case is only possible if all the components in the said circuit are *short-circuited*. If the components are not short-circuited, a practical solution is to have current and voltage sources replaced; current can be replaced by transforming a circuit into an open circuit, and voltage can be replaced by transforming a circuit into a short circuit.

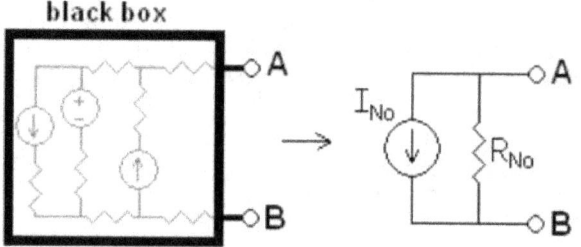

Norton's Theorem was also developed by Hands Ferdinand Mayer

The formula:

**representations: TC$_{AB}$ = total current in circuit A and circuit B

TC = total current

TC_{AB} = voltage ÷ (TC in circuit A + TC in circuit B)

VI.E. – Coulomb's Law

One of the laws that underwent heavy testing is a law that gives importance to the electrostatic relationship of all the charged elements within a circuit; this law is called *Coulomb's Law* after the scientist, Charles Agustin de Coulomb. It argues that the magnitude of interactive circuit components is in direct proportion to the *squared area* between the distances. To demonstrate a point, it shows that the primary force works along a straight line.

Moreover, Coulomb's Law requires that the placement of the charged elements is accomplished through a single medium. This causes the arrangement to eliminate any possible complications.

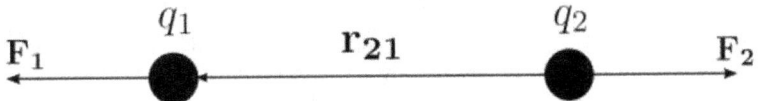

The application of Coulomb's Law is only valid if the circuit components are point charges

To make it effortless to obey Coulomb's Law, the determination of *Coulomb's constant* is recommended. Once the value of the constant has been determined, not only is it easier to adhere to the aforementioned law, you are also ensuring that the calculations for the electric current are valid.

The formula:

**representations: CC = Coulomb's Constant

ε_0 = electric constant

$CC = 1 \div 4\pi \times \varepsilon_0$

Chapter VII – Understanding Electromagnetism

A pair of loudspeakers is an example of an electronic device that works due to the *strategic positioning* of circuit elements. That, along with the integral parts called *electromagnets*; such components cause disturbance to other circuit elements. Their effect on their fellows is rather significant as they tend to make the device's cone to vibrate. As a result, sound, and in most cases, high quality sound, is produced.

For a deeper understanding of a loudspeaker, you could use a lesson in *electromagnetism*.

VII.A. – An Introduction to the Concept behind Electromagnetism

Electromagnetism is a branch of science that involves the study of electricity and magnetism. It follows that wherever there is an electric field, there is also a magnetic field. It was initially introduced and developed by *Hans Christian Orsted*. In 1802, an Italian scholar by the name of *Gian Domenico Romagnosi* also examined the field

In circuitry, understanding the concept behind electromagnetism is important due the possibly strong electromagnetic reaction of circuit components. It follows that the *EMF* (or electromagnetic force) of any circuit serves a major role in the determination of internal properties.

Chapter V, *Power Transfer at Max*, and Chapter VI, *Laws, Laws & More Laws*, contain different discussions about transferring electric current and voltage. In electromagnetism, more discussions about transfer will be tackled. This time, however, the focus is on the charged elements in a circuit (whether negatively or positively charged), as well as their responses when exposed to movement.

The primary principles of electromagnetism:

- The direction of a magnetic field dictates the direction of the current in a circuit

- The electric current of a conductor creates a magnetic field; the formation is dependent on the direction of the conductor, but is always shaped in a corresponding circle

- The magnetic poles of charged elements always come in pairs; one pole in a pair needs to be inversely proportional to the other

- Charged elements can either attract or repel; elements with different charges are attractive to each other, while those with similar charges avoid contact

- Electric current can be induced when it is moved away or toward a magnetic field

VII.B. – Gauss Who?

A notable individual in the field of electromagnetism is *Carl Friedrich Gauss.* He is a German mathematician who put together the critical electromagnetism concepts in the form of a law called *Gauss Law.* According to him, to determine the relevance of the distribution of electric charges, the circuit's electric field, in its entirety, must be evaluated.

Gauss Law allows mathematical expression with the employment of *integral* and *differential calculus,* preferably in vector form. For a beginner in circuitry, the particular technique of demonstration is advised since it grants a clear perspective of the concept of electromagnetism.

The integral formula:

**representations: Φ_E = electric field

ε_0 = electric constant

Q = total electric charge

$$\Phi_E = Q \div \varepsilon_0$$

The differential formula:

**representations: ∇ = an electric field's divergence

E = the other half of an electric field's divergence

p = electric charge density

ε_0 = electric constant

$$\nabla \times E = p \div \varepsilon_0$$

VII.C. – The Main Formulas in Electromagnetism

In a circuit's electromagnetic field, it is important to remember that charged elements tend to move radically; sometimes, predicting the direction of the electric flow is nearly impossible. While others go about in a non-linear network, many charges obey the rule of superposition since they adhere to a linear path. With the use of certain laws as basis, the relationship between the charged elements can be evaluated.

4 laws:

1. Ampere's Law – an important law whose applications include instances of a moving magnetic field; particularly, it be applied in situations involving current-carrying wires

   The formula:

   **representations: B = electromagnetic field

   DL = differential element of the electric current

   μo = permeability of o space

   I = electric current in an enclosed circuit

   $\int B \, X \, DL = \mu o \, X \, I$

2. Biot-Savart's Law – a law that is employed for the calculation of steady current in an electromagnetic field; a requirement is a constant time variable, as well as a charge that is a subject of neither a build-up or depletion

   The formula:

   **representations: B = electromagnetic field outside a circuit

   μo = permeability of o space

   I = electric current in an enclosed circuit

   R = distance from the electromagnetic field

   $B = \int (\mu o \, X \, I) \div 4\pi \, R^2$

3. Faraday's Law – like his [Michael Faraday] law of induction, this is a law that addresses the induced electromagnetic force of an object; it is strictly applicable to the charged elements within a closed circuit

   The formula:

   **representations: IEMF = induced electromagnetic field

   Dφ = difference of space

   DT = time differential

   $IEMF = - D\varphi \div DT$

4. Lorentz Force – a law that assesses a point charge due to both an electromagnetic field and an electric field

   The formula:

**representations: LF = Lorentz Force

q = total charge

EMF = electromagnetic field inside a circuit

V = velocity

B = electromagnetic field outside a circuit

$$LF = q\,[EMF + (V \times B)]$$

VII.D. – Electrodynamics & Quantum Electrodynamics

Before the 1900s, a scientist named William Gilbert addressed a proposal concerning *electricity* and *magnetism*; according to him, while both subjects can be traced to attracting and repulsing objects within a circuit, electricity and magnetism are different concepts. The key to understanding electromagnetism, therefore, is to understand the individual terms; particularly, understand their relevance and distinction.

A conflict regarding electromagnetism is that, despite agreeing to Albert Einstein's *Special Relativity*, it goes against some of the rules of mechanics; it is only dependent on the electromagnetic permeability of 0 space. It follows that in the case of moving frames, the electromagnetic field is subjected to transformation to include space.

Moreover, all electromagnetic phenomena are covered under *quantum mechanics*. This makes the electromagnetic field of a circuit accountable for the physical phenomena that are observable, especially *magnetism and electricity*.

Chapter VIII – Let's Talk Circuit Boards

Are you familiar with *crocodile clips*?

Crocodile clips are devices that can be used for the assembly of a circuit; these tools are so-named for their resemblance to the jaws of a crocodile. With them, a solid grip to connect a component is possible. If you're wondering how electrical engineers can create electrical connections without the associated dangers? Well, there's your answer.

By using *crocodile clips*, you can make a model of a working circuit. Whether be it a basic circuit, a series circuit, or a parallel circuit, you can create for an audience to analyze.

VIII.A. - Printed Circuit Boards 101

Printed circuit boards (or PCBs) are devices that enable electrical connectivity even in an "open" environment; within their system, there are resistors, inductors, transformers, capacitors, conductors, and semi-conductors. These tools support high component density. In a way, printed circuit boards are referred to as *live circuits*.

Since functioning circuit boards can be rather risky, especially when exposed to extreme environments, they are packaged accordingly. In most cases, these devices are subjected to a series of coating procedures and are dipped in acrylic, wax, polyurethane, and epoxy.

Design standards:

- Templates and card dimensions are designed according to required circuitry regulations

- Manufactured Gerber data are generated

- Design is planned thoroughly with the assistance of an *EDA or Electronic Design Automation* tool

- Traces for signals are routed

- Copper thickness and layer thickness are carefully evaluated

VIII.B. – Circuit Board Tests

To see if a circuit works, certain tests are performed. Particularly, it is determined whether or not it is functional and can perform desired tasks. Along with its capacitors, resistors, transformers, and other components, it is analyzed for opens and shorts.

Objectives of testing methods:

- To detect flaws

- To detect error-free operations of each of its components

- To determine system stability

- To evaluate whether it is fit for use

- To evaluate safety issues

- To verify test systems

Example testing methods:

- Analog tests

- Contact tests

- Contact tests

- Electrolytic capacitor tests

- Flash tests

- Powered digital tests

- Short tests

VIII.C. – Let's Learn to Prototype

Prototyping is the ability to put a particular idea to test by preparing a model from which other circuits are developed. Especially when the circuit in subject involves a complex system or expensive components, a prototype is initially designed.

If the circuit creator is unsure of how a particular circuit will function, his best bet is to create a *prototype*. This way, he can evaluate his creation and see how it can be improved. If it's not yielding his desired results, he can modify the placements of each of its components until he achieves a necessary output. Otherwise, he can proceed to the actual circuit-making process.

VIII.D. - The Art of Bread-Boarding

Breadboards allow the creation of a circuit prototype, which is the reason why these tools are ideal for beginners. *Bread-boarding*, therefore, is the process of creating a circuit prototype in a board to resemble the operations of a real circuit. Its history can be traced to the time that electrical enthusiasts would use a literal breadboard (i.e. the board used for slicing bread).

According to experts, it's recommended to learn and understand bread-boarding prior to making your first-ever circuit. It's advised to be familiar with each of its components, as well as how it works.

Breadboard components:

- Chips – these are *legs* that come out of both sides of a breadboards; these components fit perfectly and serve as connectors of different parts

- Posts – these are components that enable connections from power sources

- Power rails – these are vertical metal rows strips that are adjacent to terminals; through these components, easy access to a power source can be provided

- Power supplies – these are components that enable the supplementation of a wide range of electric current and voltage levels

- Terminals – these are horizontal metal row strips that are adjacent to power rails; through these components, wires are allowed to be inserted, then, be held intact

VIII.E. – Essential Skills in Circuitry

After understanding *bread-boarding*, you're almost ready for some hands-on circuit lessons. First, however, you should adopt a certain set of skills; while bread-boarding is helpful, it's only practice (i.e. for instance, there is no soldering involved). Since you'll be making an actual functioning circuit, it's time to get your hands dirty.

5 skills required in circuitry:

1. Stripping (wire) – it is a skill that promotes secure electrical connections; it involves the knowledge on various types of wires, thickness of wires, and how to maintain a solid grip; since exposure to electric current is a risk, it is best to check out the appropriate tools for wires

2. Drilling – it is a skill that focuses on the proper drilling (of holes) in a circuit, then, making sure that electrical connections can remain intact; it is a slow process that requires practice with accuracy and precision

3. How to test batteries – it is a skill that requires testing the capacity of a battery; particularly, it involves measuring the current load in open and short circuits

4. How to use a glue gun – it is a skill that takes advantage of a glue gun's conductive property; it teaches how to insulate and how to set a semi-permanent coating; especially when there is a need to strengthen the joints of a circuit, it comes in handy

5. How to use liquid electrical coating – it is a skill that focuses on the ability to apply liquid electrical coating where conventional electrical tape is likely to fail; it requires precaution since about 30% of the parts in the liquid coating are quite volatile

VIII.F. – The Secrets of a Solder

Soldering is a process that involves (at least) 2 metals (or any conductive material); the 2 metals are joined by flowing and melting. In the industry of circuitry, it is one of the most fundamental methods in circuit creation; it allows independent components to work as one.

A technique of good soldering revolves around the knowledge of the amount of heat that is applied. For *basic soldering*, *361F* is the temperature to consider and for *advanced soldering*, the goal is to arrive at a temperature of somewhere between *361F to 419F*.

In *metallurgical engineering*, a term named *flux* is commonly used; in circuit engineering, as soon as the circuit creation process commences, it will be introduced, too. It is cleaning agent and a flowing material that facilitates the soldering process. Should there be invisible impurities (e.g. oil, dirt, etc.), they will be removed for the purpose of not risking the integrity of a circuit.

Moreover, it is important to note that in soldering, the proper application of flux is suggested. Improper methods can result to joint failure. The system's damage (due to incorrect flux application) may not be obvious at first, but gradually, it is capable of corrosion and rendering a circuit useless.

Chapter IX – Sufficient Safety

On a sunny day, try heading outside of your house, then, checking out the electrical connections (wires) from one post to another. More likely, one, two, or even a flock of birds are calmly resting on power lines.

Do you ever wonder why they don't get shocked?

No, birds are normal creatures; they do not possess extra-special powers. In their heads, they understand that it's a must not to step on open electrical networks, which is one of the things that need to be covered for beginners in circuit engineering.

IX.A. – The Lack of Electrical Safety Courses

While some *circuit engineering professors* teach enough lessons about *electrical safety*, others are quite behind on the area. They assume that practical knowledge, as well as a general "be careful" would suffice; since the beginners in circuit engineering are frequent exposure to electrical devices, no, the statement won't cut it. Therefore, it is important to stress out the need to be careful especially during their first hands-on activity.

Electrical safety tips:

- Treat each electrical device as if live current is running inside it (regardless if you're aware that there isn't)

- Overloading sockets for a circuit board? Not a good idea

- To cut off running current, add a residual current

- Always disconnect (and not just turn off) electrical devices when working on a component

- Always turn off any electrical device if not in use

- Regularly check the conditions of sockets and plugs

- Practice extreme caution when dealing with liquids and electric current

- Make sure your hands are dry

- Avoid octopus connections (i.e. devices that enable multiple plugs or sockets)

- Keep electrical wires and cords tucked neatly

- Wear the proper attire when creating a circuit; put on non-conductive gloves and footwear with insulated soles

IX.B. - The Importance of Hands-on Circuit Lessons

Hands-on learning is important for a beginner in circuits. It gives light to the fact that in circuit engineering, it is more about actual field work. Once the concepts are understood, it's best to move on to creating functional circuit boards.

A common problem that is encountered by beginners? Sweaty hands. Even in an air-conditioned laboratory, there are people who have to deal with sweaty hands when handling circuits. Alongside, they have to deal with the need of maintaining a strong and solid grip on various tools. Although this may be a concern, it's one that requires practice.

IX.C. – The 80-20 Rule of Circuit Safety

In circuitry, there is a rule that focuses on the installments of transmitters in hazardous areas for devices with a circuit; it is called the *80-20 Rule of Circuit Safety*. According to the rule, it is recommended to take extra precaution when dealing with particular portions of an electronic device; there are some parts that increase the risk of electrical shock when *touched* or *moved*.

Hazardous areas of a circuit (that require extra precaution):

- Point with high impedance

- Point with high resistance

- Area near the voltage source

- Area near an opening

- Area with conductive elements

IX.D. – Troubleshooting Concerns

In the event that a device with a circuit is not function correctly, it is recommended to have it evaluated accordingly. While a beginner may handle the task of checking, a professional in circuitry is the one who is advised to assess the system. Since an expert is already familiar with different circuit components, and he knows exactly where to look for possible faults, he is more qualified for the job.

Troubleshooting tips:

- Determine whether the connections are secure

- Determine whether the wires are correctly connected to one another

- Check for circuit components that seem out of place

- Check for circuit components that may be larger in size

Chapter X – Here's a Multimeter for You!

An important and indispensible tool in circuitry is a *multimeter*.

Imagine a situation when a circuit project was presented to you. Since you were requested to detail accurate measurements for its electric current, voltage, and resistance, you go to the nearest equipment laboratory to borrow a stack of devices for assistance.

Now, imagine the same situation, but this time, you have this tool; instead of having to go to the equipment laboratory to borrow a stack of devices to measure the electric current, voltage, and resistance, you turn to that tool. This particular time, you have a multimeter.

If you're not familiar with the operations of a multimeter, let it be your first job prior to creating a circuit project. It's actually quite easy to use. So long as you are attentive to instructions, and you know its components, certain restriction, and the technique to maximizing its functions, you're set.

X.A. - What's a Multimeter?

A multimeter, also called a multitester, *Volt-Ohm millimeter*, and *Volt-Ohm meter*, is an electronic device that measures the electric current in a circuit; it is capable of measuring voltage, resistance, and a variety of other units in a circuit's system. To know how to use the device, it's important to be familiar with each of its components; if you can understand how the components work, you can also understand the operations of the entire device.

Parts of a multimeter:

1. The selection knob – it enables a user to choose a particular setting that is subject for measurement

2. The display meter – it enables the display of the measured reading; it can contain up to 4 digits, as well as a negative sign

3. The probes – these are plugged into a multimeter that can interpret and convert measurements from a device into a multimeter; usually, these come in a pair of red and black probes

Types of probes:

I.   IC hook

II.   Alligator clips

III.   Test probes

X.B. - A Multimeter in the Works

Inarguably, a multimeter is one of the most useful tools in circuitry and one of the tools that can make the task of building circuits easy. It can be used to measure *any object or device* that contains a circuit and an electric current. Using it, however, comes the hard part; in fact, in the world of circuits and electronics, it earned the title as one of the most challenging jobs. It may be effortless to get a reading, but getting an accurate reading, then interpreting the reading is another story.

How to use a multimeter to measure (a basic example):

1. Prepare an AAA battery.

2. Plug the black probe of a multimeter to the negative side (i.e. the side with a "-") of the AAA battery.

3. Plug the red probe of a multimeter to the positive side of the AAA battery.

4. Check the display meter; as recommended, check the meter twice.

5. List down the reading and begin the interpretation.

X.C. – Resolution, Accuracy & Input Impedance

The smallest part of a multimeter's scale is called as the *resolution*. It is responsible for achieving an accurate reading and interpretation. In many multimeter kinds, especially digital ones, it can be configured or *calibrated*. And, as the rules go, a device with low resolution doesn't require much completion time; a device with high resolution can require a demanding processing time.

Meanwhile, the *accuracy* of a multimeter refers to an error in the measurement of an electric current, in comparison to a perfect reading. It is relative to the device's resolution since resolution may not be calibrated accordingly if the absolute accuracy level is questionable. Therefore, to determine the total accuracy of a multimeter, its relative accuracy should be added to its absolute accuracy.

Formula for the computation of total accuracy:

**representations: TA = total accuracy

RA = relative accuracy

AA = absolute accuracy

$TA = RA + AA$

When talking about a multimeter's resolution and accuracy, *input impedance* is a set of terms that needs to be acknowledged, too. This is due to the device's inability to achieve accurate readings when it is not set accordingly. Especially for the measurement of a circuit's voltage, its input impedance has to be calibrated high (i.e. higher than a circuit's voltage) so the operation remains smoothly.

X.D. – Safety Concerns

Groups that are in charge of the manufacture of multimeters, as well as the authorities that promote the safe use of such devices have set safety standards. This is to emphasize the importance of the right employment of the electrical tools. Although the method for use can be rather straightforward, reckless habits can result to problems. Alongside the possibility of inaccurate readings, it can cause harm to the individual that is handling the devices.

Categories of safety standards:

- Category 1 – applicable to the employment of a multimeter, circuit, or any electronic equipment with a distance near main connections

- Category 2 – applicable to the employment of a multimeter, circuit, or any electronic equipment with a distance somewhere near the first phase of main connections

- Category 3 –applicable to the employment of a multimeter, circuit, or any electronic equipment with a distance near permanently installed loads

- Category 4 – applicable to the employment of a multimeter, circuit, or any electronic equipment with a distance near faulty current levels that can be quite high

Chapter XI - DIY Circuits: Simple Projects

When you're into circuits, as well as *electronics and electrical engineering*, you need to step up your game when it comes to building items. Isn't the main reason for learning the branch of electrical engineering for the creation of circuits, then, putting them to good use?

In the event that your first circuit project wouldn't turn out as desired, try not to get discouraged easily, and instead, give matters another go. Not getting the results you wanted maybe a bit of a downer, but eventually, the odds will be in your favor; look at the setback as an opportunity for learning. If your heart's into circuits, you'll soon get the hang of how things are done.

XI.A. – Common Tools in Circuitry

For the creation of a circuit, you can make use of just about any tool you come across; if you find equipment that will make you accomplish a task easier, then, maybe you should put it in your arsenal. You *can* use just any tool, yes, but, doing so is not advised. It is best to choose the right set (i.e. a set of tools with a non-conductive handle) to not put your safety at risk.

Common tools:

- Crimper

- Cutters (e.g. cable cutter, electrical cutter, etc.)

- Extraction tool

- Glue gun

- Non-metallic tweezers

- Pliers

- Screwdriver

- Soldering gun

- Wire-wrapping tools

XI.B. - Beginner Circuit Projects

Building your first circuit can be a challenge; since you're still a beginner, you may end up making mistakes. Maybe your circuit won't end up as functional as desired, despite having followed your understanding of a series of procedures. In such a case, this is where you have to tweak your work; be diligent in figuring out where you went wrong. If you're uncertain

Circuit Engineering: The Beginner's Guide to Electronic Circuits, Semi-Conductors, Circuit Boards and Basic Electronics

of your actions' impact, don't worry too much. The important part is to begin; you can, then, figure out the rest along the way.

7 easy circuit projects (derived from http://www.instructables.com):

- Project # 1 – Static Electricity Analyzer

    The Static Electricity Analyzer can detect nearby static electricity; to indicate that static electricity is present, its LED component glows. Apart from a detector, it can be used to analyze the electricity in its surroundings. It is an extremely sensitive device since it can even detect nearby hand movement without touching the antenna.

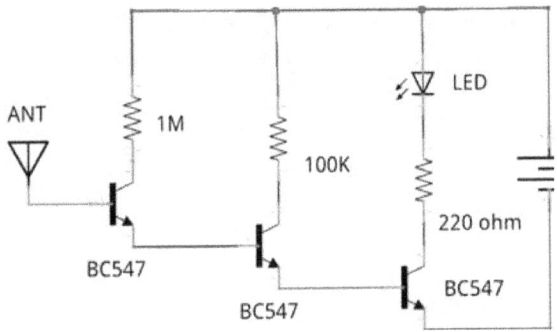

Materials:

- Around 10 pieces wire
- 1 piece LED
- 1 piece static electricity antenna
- 1 piece 100K resistor
- 1 piece 1M resistor
- 3 pieces 2n222 transistor

Procedures (as shown in the layout above):

1. To the left side of the circuit board, connect the 1M resistor to 2n222 transistor.
2. Beside it, attach the wire, then, attach the static electricity antenna.

51

3. On the bottom, attach the wire, then, attach the other 2n222 transistor; next to it, attach the wire, and finally, attach the third 2n222 transistor.

4. Place the 100K resistor adjacent to the second 2n222 transistor.

5. Place the LED adjacent to the third 2n222 resistor.

6. Establish connection to a power supply.

- Project # 2 – Dark LED Light

The Dark LED Light is a circuit project that can detect darkness. It follows that when insufficient light is supplied, an IC timer is alerted; consequently, a high output is produced and LED light will be switched on. The idea behind it is similar to that of a street light that automatically turns on once it detects that it's already evening.

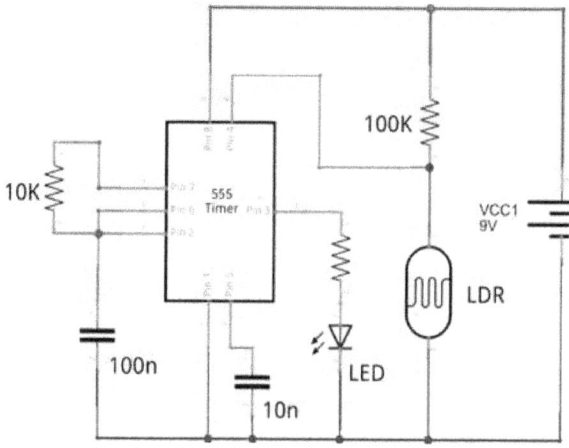

Materials:

- Around 10 pieces wire

- 1 piece LED

- 1 piece LDR

- 1 piece 10nf capacitor
- 1 piece 100nf capacitor
- 1 piece 10K resistor
- 1 piece 100K resistor
- 1 piece 555 IC timer

Procedures (as shown in the layout above):

1. On top of the LDR, attach the 100K resistor.

2. On the bottom of the LDR, create a connection to the LED.

3. To the left of the LED, create a connection to the 10nf capacitor.

4. To the left of the 10nf capacitor, create a connection to the 100nf capacitor.

5. Next to the 100nf capacitor, create a connection to the 10K resistor.

6. Place the 555 IC timer between all the connections; establish the main connection by attaching the other ends of the LED, LDR, capacitors, and resistors to the 555 IC timer (directly or with a wire).

7. Establish connection to a power supply.

- Project # 3 – The Ticking Bomb

The Ticking Bomb is meant for creating a ticking sound that resembles a bomb. Once it is turned on, it produces sound that is adjustable, but is modified to 1 tick per second.

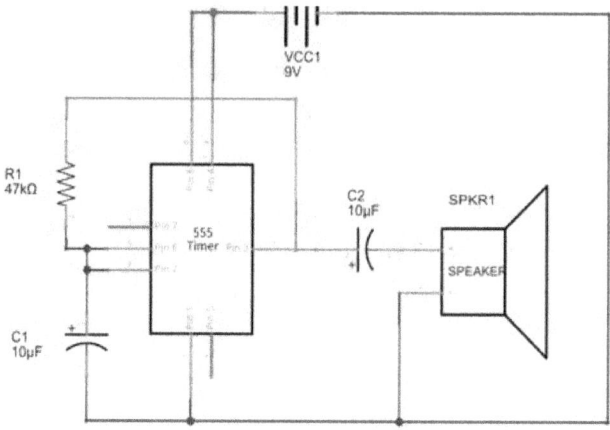

Materials:

- Around 10 pieces wire
- 2 pieces 10uf capacitors
- 1 piece 555 IC timer
- 1 piece 47K resistor
- 1 piece 8 ohm speaker

Procedures (as shown in the layout above):

1. Make the 555 IC timer as the central component; to its left, create a connection to the 47K resistor.

2. On the bottom, create a connection to the first 10uf capacitor; then, create a connection from the 10uf capacitor to the 47K resistor.

3. To the right of the 555 IC timer, create a connection to the second 10uf capacitor.

4. From the second 10uf capacitor, create a connection to the 8 ohm speaker.

5. Establish connection to a power supply.

- Project # 4 – The Remote Tester

The Remote Tester, as its name suggests, is a circuit that checks whether or not a remote control is working. Behind it, the idea is focused on the sufficient amount of signals that the IR receiver is getting. If it receives enough signals, the LED lights up, which means that a particular remote control is functioning; conversely, if the LED remains as is, it is an indication of a faulty component in the device.

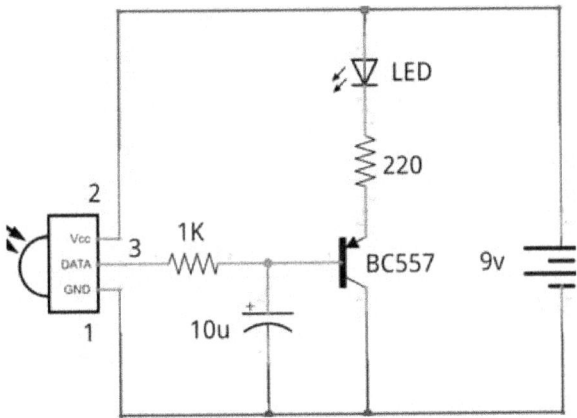

Materials:

- Around 10 pieces wire

- 1 piece LED

- 1 piece IR receiver

- 1 piece 1K resistor

- 1 piece bc557 transistor

- 1 piece 10uf capacitor

Procedures (as shown in the layout above):

1. On top of the bc557 transistor, create a connection to the LED.

2. Adjacent of the bc557 transistor (to the left), create a direct path to the 1K resistor, then, to the IR receiver.

3. On the bottom, create a connection to the 10uf capacitor.

4. Connect the 10uf capacitor (directly or with wire) to close the connection.

5. Establish connection to a power supply.

- Project # 5 – The Bell Experiment

The Bell Experiment is a basic project that produces a musical sound; the result is a device that may be similar to a doorbell. It works according to each of its components; if the resistor, transistor, and IC are triggered, the bunch sends a signal to the speaker, which will then, create the sound.

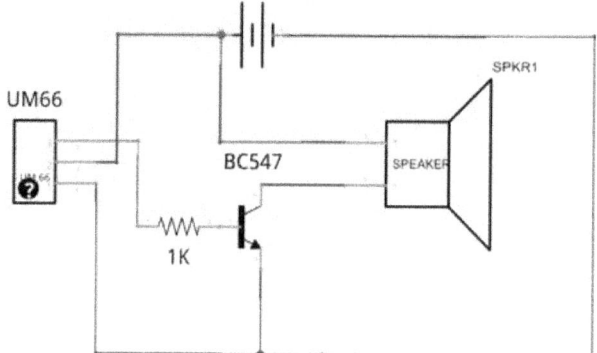

Materials:

- Around 10 pieces wire

- 1 piece 1K resistor

- 1 piece 2n222 transistor

- 1 piece UM66 IC

- 1 piece 8 ohm speaker

Circuit Engineering: The Beginner's Guide to Electronic Circuits, Semi-Conductors, Circuit Boards and Basic Electronics

Procedures (as shown in the layout above):

1. On one end of the 8 ohm speaker, create a connection to the 2n222 transistor.

2. On another end, create a connection to the 1K resistor.

3. From the 1K resistor, create a connection to the UM66 IC; make sure that the UM66 is parallel to the 8 ohm speaker.

4. Connect the UM66 (directly or with wire) to the 2n222 transistor to close the circuit.

5. Establish connection to a power supply.

- Project # 6 – The LED That Fades

The LED That Fades is a project that produces and sometimes, blinking lights. It operates according to the weakness or strength of the signals that are interpreted by each of its components. If its IC timer, transistor, resistor, and capacitor receive strong signals; the LED will glow; conversely, if they receive weak signals, the light starts to fade. In the event that the signals that are submitted are unstable (i.e. they alternate between weak or strong in a few minutes' time), the blinking pattern comes in.

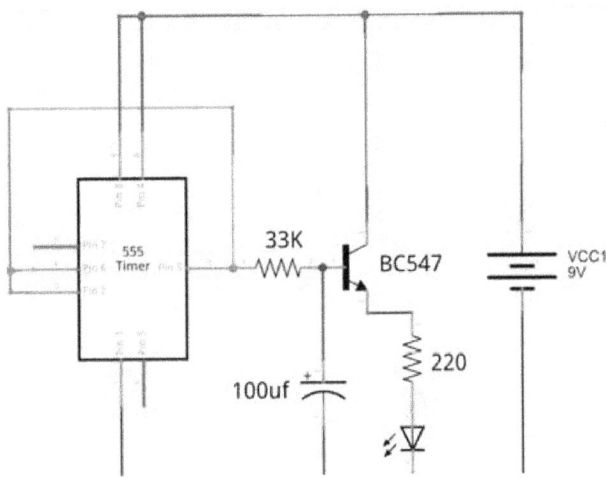

Circuit Engineering: The Beginner's Guide to Electronic Circuits, Semi-Conductors, Circuit Boards and Basic Electronics

Materials:

- Around 10 pieces wire
- 1 piece LED
- 1 piece 555 IC timer
- 1 piece bc547 transistor
- 1 piece 33K resistor
- 1 piece 220 ohm resistor
- 1 piece 100uf capacitor

Procedures (as shown in the layout above):

1. To the bottom of the bc547 transistor, create a connection to the LED.

2. Parallel to the LED, create a connection to the 100uf capacitor.

3. On top of the 100uf 33K resistor, create a connection to the 220 ohm resistor.

4. Next to the 220 ohm resistor, create a connection to the 555 IC timer.

5. Create a connection around the 555 IC timer for a closed circuit.

6. Establish connection to a power supply.

- Project # 7 – The LED with Activated Light

The LED with Activated Light is a basic project for beginners in circuit engineering; it is meant for a clearer understanding of the concept of resistance. The LED lights up with the application of sufficient resistance; if resistance levels are insufficient, the light won't be activated.

Circuit Engineering: The Beginner's Guide to Electronic Circuits, Semi-Conductors, Circuit Boards and Basic Electronics

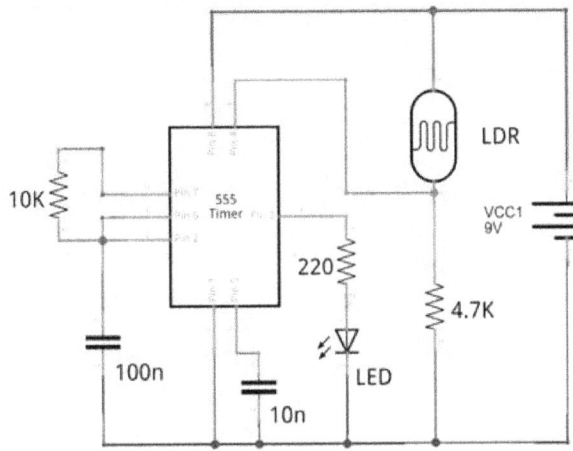

Materials:

- Around 10 pieces wire
- 1 piece LED
- 1 piece LDR
- 1 piece 10nf capacitor
- 1 piece 100nf capacitor
- 1 piece 10K resistor
- 1 piece 4.7K resistor
- 1 piece 555 IC timer
- 1 piece 220 ohm resistor

Procedures (as shown in the layout above):

1. To the bottom of the LDR, create a connection to the 4.7K resistor.

2. Parallel to the 4.7K resistor connection, create a new connection to the LED, followed by the 22 ohm resistor on top.

3. From the 22 ohm resistor, create a connection to the 555 IC timer.

4. To the bottom of the 555 IC timer, and parallel to the LED connection, create a new connection to the 10nf capacitor.

5. To the left of the 10nf capacitor, create a connection to the 100nf capacitor.

6. On top of the 100nf capacitor, create a connection to the 10K resistor.

7. Create a connection to close the circuit.

8. Establish connection to a power supply.

Chapter XII – Making Your Way to Circuit Design: PCB Layouts & Schematic Diagrams

If you are familiar with each of its components, you may have a good grasp of how an electronic circuit works; granted that the integral parts (e.g. power source, conductor, resistor, diode, etc.) are in place, and you were introduced to different techniques on establishing connections properly, there's a high chance that any circuit project you take on can be a success. Once an opening for a power source is identified, and the stability of an arrangement for the rest in a series is achieved, you're set.

However, if given the opportunity to make a particular circuit more functional, wouldn't you take it? Instead of settling for a random circuit design, why not have it modified for superior performance?

XII.A. - Circuit Design 101

Ever wonder why circuit engineers are paid highly and (depending on their chosen career path) are presented different career advancement opportunities (e.g. work as computer engineers, robotics specialists jobs, etc.)?

The responsibilities of circuit engineers are rather demanding; they need to be mentally tough and be open to different challenges. They spend hours and hours racking their brains out to determine how systems can be made more efficient. Chances are, there's always a way; it's up to them to look for one. When a particular project asks for it, it's mandatory for them to let their creative juices flow to arrive at a compatible solution. Otherwise, their built circuits may stop functioning eventually.

As mentioned, it's practical to design a circuit accordingly; especially with their employment for the fields of *communication, navigation, telecommunication, travel*, and other industries, their designs should consider a particular purpose. Apart from the objective of meeting different requirements, there needs to be a strategy since a more functional circuit comes with higher quality; it can make a project consume a reduced number of resources, too.

Reasons why there are various circuit designs:

- Improvement of a circuit's efficiency

- Improvement of a circuit's size and weight

- Guarantee a circuit's durability for a set period

- Guarantee a circuit is safe to use

XII.B. – The Design Process

Many times, a PCB layout (as discussed in chapter VII, *Let's Talk about Circuit Boards*) and a schematic diagram are often interchanged in circuitry; it has led others, especially the novices in circuit engineering, to believe that they are one and the same. In a way, since they are both presentations of a circuit's system, they may seem similar. However, upon closer inspection, they are not. It is, therefore, essential in the circuit design process that the difference is acknowledged.

A *PCB layout* is a physical representation or a *real model* of a circuit. It presents all of the electrical components that are included; it also details which of the components are active and which ones are passive. Although it can show an actual working circuit, understanding the functions of each of its parts can be tedious.

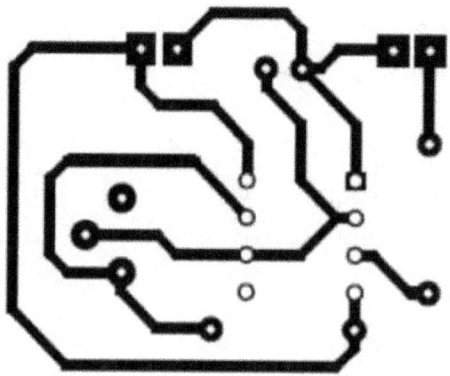

A PCB layout is an actual circuit model

On the other hand, a *schematic diagram*, also called simply as a circuit diagram, is a descriptive outline of a circuit's system; it is the standard and less costly way of circuit representation. It is like a PCB layout that shows both of the active and passive components in a circuit; unlike a PCB layout, however, its presentations can be easily understood.

Circuit Engineering: The Beginner's Guide to Electronic Circuits, Semi-Conductors, Circuit Boards and Basic Electronics

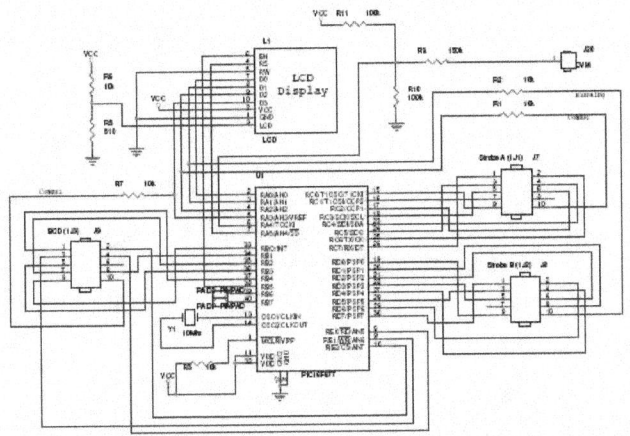

A schematic diagram can include important notes for the improvement of a circuit's design

Stages of circuit design:

- Meeting specific requirements

- The creation of a circuit diagram

- Building a breadboard, or a PCB layout

- The presentation of each circuit component (for professional evaluation)

- Applying the results of evaluation

- Testing (and retesting)

- Getting approval from professionals

Circuit design tips for beginners (preparation process):

- Categorize the components

- Construct a PCB layout, as well as a schematic diagram

- Determine the compatibility of each circuit component

Circuit design tips for beginners (circuit-building process):

- Always avoid cold solder joints

- Always separate power controls and other connections

- Always separate analog and digital components

- Create traces should hard-to-find components be included

- Remember to always make integral nodes accessible

- Solder the components systematically; solder small components first, then, solder larger components next

- Strategize the spaces you allow between components

- Take note of any heat spots

XII.C. - Circuit Symbols

On the process of building a circuit and modifying its design, an understanding of different circuit symbols is important. Descriptions of circuit components can be put in simple words; it is called a *verbal description*. However, since a thorough approach is recommended in many circuitry lessons, a visual representation can enable you to understand the electric flow in a circuit.

Here's a good comparison:

Verbal description	Visual representation
Circuit # 1 contains a light bulb, as well as D-cell battery as power source.	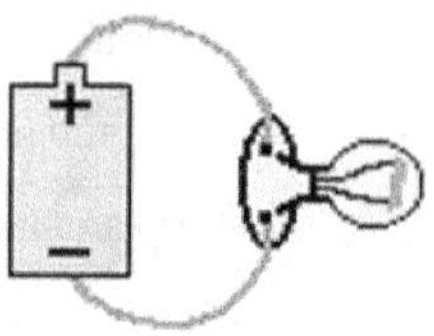

Moreover, circuit symbols for a visual representation of a circuit is preferred over a verbal descriptions since they may be less complicated to understand. Especially if the particular circuit is quite advanced, a description containing words can be challenging to use as basis. If a visual model is employed, however, you can simply focus on connections, instead of the interpretation of a worded description; there is a better chance of building an impressive circuit design successfully. Especially if you have

64

plans of advancing your place in circuit engineering by soon moving forward from a novice to a circuit expert, *memorizing* the different circuit symbols is a must.

List of basic symbols:

Symbol name	Symbol
AC	
DC	
Capacitor	
Resistor	
Fuse	
Diode	
LED	
Regulator	
Transistor	

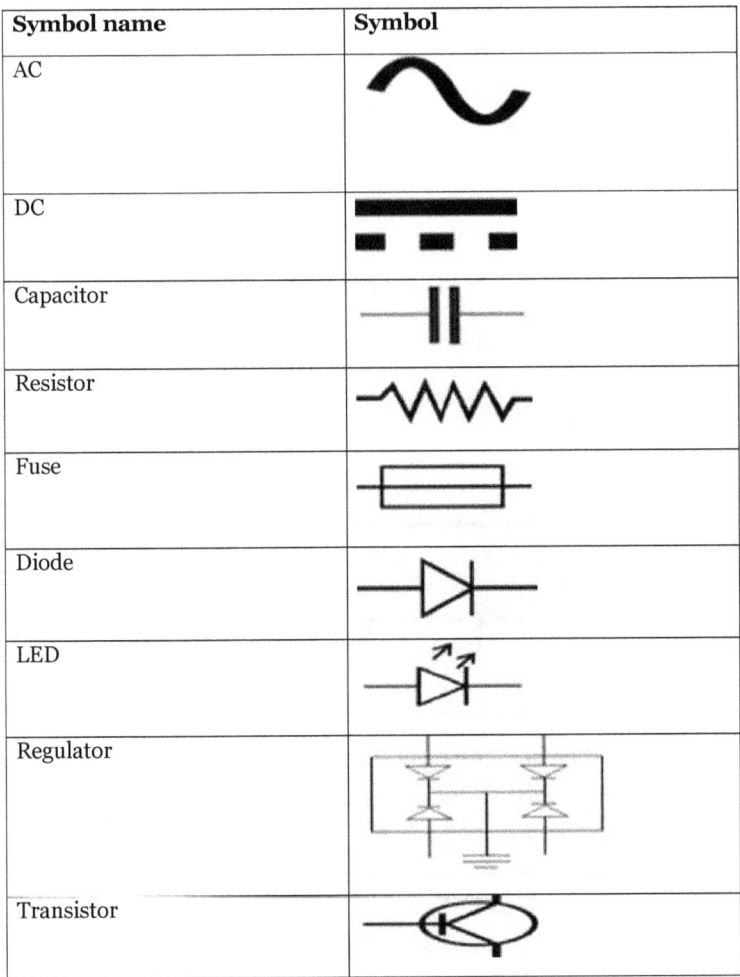

List of basic capacitor variation symbols:

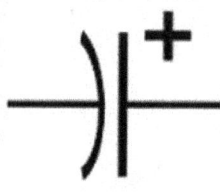

-

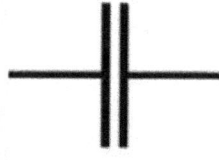

-

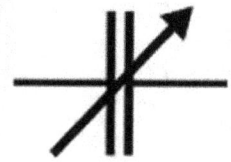

-

List of basic diode variation symbols:

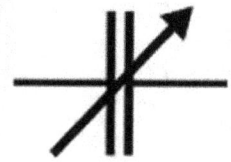

-

Anode ▷|◁ Cathode

-

Anode ▷|| Cathode

-

Anode ▷| Cathode

-

List of basic switches variation symbols:

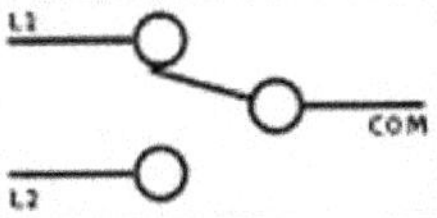

L1 ⟋ COM
L2

-

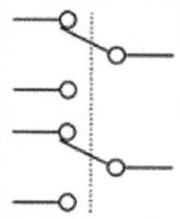

•

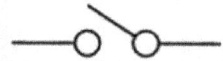

•

List of basic transistor variation symbols:

•

•

-

-

XII.D. – Factors to Consider

A strategic circuit design is important since a circuit may not function at all in the absence of a good plan; if the placement of each of its components wasn't accomplished according, you may have just wasted resources. You may have the correct components but if they are not designed properly, a circuit is far from working. Besides, at the realization that you may not be design a circuit's system according to requirements, you wouldn't want to put the circuit-building project to an end at the middle of the process, right?

Moreover, the proper design structure of a circuit needs to be prioritized. Regardless if it is not meant to impress another, it should still be designed accordingly as to be a stepping stone in the possibility of handling more projects eventually. With the knowledge of how to design a circuit properly as a beginner comes the effortless knowledge on the construction of more advanced circuits eventually.

What to keep in mind:

- Time availability (of the creator) – although it can be completed quickly, a circuit project shouldn't be rushed; the project creator should be available for a set period so he can concentrate properly on a circuit. Especially if he has hardly any time to spare, it's suggested that he either determines a way to find additional time, or takes on a different project

- Difficulty level – the assessment of a circuit project is important prior to acceptance. While there are simple ones, some projects are rather complicated. If a creator is certain that he can't follow through the procedures of a particular project, he may want to consider other undertakings

- Total cost of the project – one reason that the creation of a schematic diagram and other preparations are suggested (prior to taking on a circuit project) is due to practicality regarding the overall expenses. As much as possible, the creator should consider his ability to fund a project. Especially if he plans on using expensive circuit components, he needs to think about whether he can afford the project until completion.

- The desired function – identify the reason why you took on the project. Is it for experimentation and personal use? Or, was it a request from a company? Especially if the purpose is for a commercial company, design a circuit to achieve first-rate quality.

XII.E. - Documentation, Verification & Testing

A primary benefit of the *documentation* process is its advantage of letting you identify possible errors in putting together a circuit. If you weren't successful at the end stage of your project, you can figure out exactly where you went wrong; in favor of having to disregard your current progress and starting from scratch the second time around, you can simply have certain methods modified (for improvement).

Verification, as defined in circuitry, is the process of thoroughly evaluating each of a circuit's components, as well as each of the stages in the making. The objective is to determine whether the process has been adhered to correctly. Usually, this can be a time-consuming process, but since it is intend to ensure quality, it is an important one.

In a way, the real-world version of the verification process is *testing*. Prior to a circuit's launch and its exposure to different commercial industries, it is tested in research laboratories. Like verification, it can also be a time-consuming process, and since it is the last process before a circuit is passed on to another source, it can be a labor-intensive series.

Chapter XIII – A Way Is through EDA

Are you familiar with *FABS* or *Semi-conductor fabrication facilities?*

FABS are places where circuit designs, even complex ones, can be put together; they are usually extensive since they are meant to accommodate bulk circuit projects. Inside them are various equipment for *EDA*, which is why circuit-production is rapid.

With the productivity level of FABS, semi-conductors are continuously becoming popular. And, since semi-conductors are components in a circuit, circuits grew in popularity, too.

XIII.A. - What Is EDA?

EDA, also known as *Electronic Design Automation* is a class of software equipment especially for designing a circuit or different electronic devices. It is categorized along with other modern tools used in circuitry; it eliminates the need to design a circuit by hand, which minimizes simple and critical errors, repair glitches, and suggest possible improvements.

Moreover, it's due to EDA that circuit creation and circuit production became faster. If you're concerned regarding the quality since the process is done in an automated fashion, be confident that each design underwent thorough analysis. Remember, the software equipment's design was specifically meant to address a circuit's system.

XIII.B. – Design Flows

In the world of circuitry, it's important to be informed of the concepts of EDA; it privileges a circuit creator with a good grasp of the proper way to build circuits. The employees of grand electronics companies such as *Intel, Hewlett Packard,* and *Valid Logic Systems* are all trained to have impressive knowledge of the tools, despite already having gone through years of education.

Through the years, EDA tools have undergone continuous development. Initially, the concepts were rather limited; in the *Age of EDA Invention,* the focus was merely on *routing, logic synthesis, static analysis,* and *placement.* Eventually, in the *Age of EDA Implementation,* these concepts were subjected to major improvements; more *sophisticated and advanced algorithm* for the EDA tools were considered. Come the *Age of EDA Integration,* they were further studied and were designed to cater to integrated environments.

Now, you may be wondering why there is a need to build circuits manually when there are EDA tools. Well, although the process can be accomplished easier and faster, sometimes, bringing the expertise of a circuit engineer on the table is recommended; this way, he can thoroughly evaluate the quality of a circuit's system.

Additionally, the operators of EDA tools are selected carefully; usually, they are the ones with a background in the operations of EDA tools and similar equipment. This is due to the expenses, as well as the possible complications, involved in using such devices. In the event that a need for troubleshooting arises, it's best to have somebody who is knowledgeable in circuit designs on standby.

Primary focuses of EDA:

- The presentation of schematic-driven layout

- Advanced and logical interpretation of each circuit component

- Hardware and transistor simulation

XIII.C. - Circuit-Level Optimization

The circuit-level optimization or *power optimization* of EDA tools refers to the various techniques that are employed for the reduction of the total power in a circuit. Although the intent is to modify a circuit's setup economically, the processes shouldn't compromise the product's overall quality. With the main goal of enhancing a circuit's efficiency, the other objectives are to increase speed, eliminate possible leakage, enhance power distribution, and improve functionality.

Circuit-level optimization techniques:

- Modification of voltage scales, thresholds, variables, blocks, supplies, and other voltage-related concerns

- Modification and re-sizing of transistors

- Modification of logic styles

- Re-routing of networks

XIII.D. - Interpretation, Analysis & Verification

Although the automation process of EDA tools may eliminate errors in a circuit's system, a product remains a subject of various testing methods. This is due to the industry of circuitry's insistence on guaranteeing the flawless networks of circuit systems. In the aim of making sure that a circuit's functions are in tune with specific requirements, it is interpreted, analyzed, and verified.

In the event when a circuit (that was created with EDA tools) hadn't undergone tests, there is a likelihood that it won't function for a specific purpose. Especially if the project will be employed for commercial agenda, the industry experts are not granting permission to distribute "unevaluated" circuit systems.

Circuit evaluation methods:

- Assessment of a circuit's maintenance requirements

- Inspection of desired and undesired effects (in relation to mathematical logic)

- Inspection of a circuit's functionality

- Inspection of a circuit's stability

- Verification of physical components

- Verification of static timing

Chapter XIV – The Other Way around: Reverse Engineering

One of the coolest ways of understanding circuits is to disassemble all of its components one by one. First, make sure that it is not connected to any power source. After putting on protective gloves, begin taking apart a component; then, determine its function. Repeat the procedures until you know the purpose of the parts and the importance of having them work as a network.

Now, put them all back together; make sure the circuit works.

Such a process is called *reverse engineering*.

XIV.A. - An Introduction to Reverse Engineering

Reverse engineering, also called *deconstruction, reversing, backwards engineering*, and *back engineering* is the process of extracting knowledge from any device or electronic equipment. It involves the need to deconstruct an entire circuit for further analysis of its components. Although the methods that will be employed are opposite to the process of creation, it is considered as a practical approach of learning circuitry.

Moreover, reverse engineering gives light to Aristotle's concept that the key to understanding the operations of a device is studying each one of its parts. Back in the days, the field is limited to its employment on a circuit and other electronic equipment; now, in the modern world, the application of a "reverse" methodology includes almost anything – from children's toys and household appliances to neuroscience, computer programming, and DNA.

XIV.B. - Reasons for Reversing

One of the privileges that reverse engineering can grant to a circuit creator is the guarantee that a device is made accordingly. In the event that he chooses to base his circuit project on another circuit that was created with *questionable privacy*, he can determine any unethical practices. If he takes apart all of its components, he has the chance to have an insider look; by then, he can confirm whether unregulated modifications are in place.

Why use reverse engineering techniques for circuit-creation:

- Documentation purposes – especially in the case of shortcomings and low-quality circuit documentation, the diagnosis of a circuit is necessary to present new information

74

- Interfacing – with reverse engineering techniques can be subjected to interfacing; it can be evaluated accordingly, regarding its compatibility with another circuit

- Bug fixing – if there are critical faults in a circuit's design, it can be recognized better with a closer look at each of its components; instead of resorting to assumptions, a circuit creator can identify which part of his project needs modification

- Advanced technical information – advanced technical information is rewarded when opening up a circuit; especially for beginners, reverse engineering is a practical way of learning about the straightforward details within the system

- Incorporation of a new functionality – reverse engineering can allow the incorporation of a feature in a circuit; rather than design another circuit, a circuit creator can simply make modifications on his current project

- Modernization – reversing is beneficial in circuitry since it can be used to modernize a project; for instance, if most modern devices are popular in the market due to an innovative feature, a circuit creator can employ reverse engineering to add the same feature to his circuit

- Guarantee product security – if a circuit creator is unsure of the security of his project, reverse engineering is a way for him to determine certain concerns; he can check the specifications of each component and guarantee that none poses safety risks when put to use

XIV.C. – Is Reverse Engineering Similar to Hacking?

Since it is the method for the extraction of information on a circuit that can't be retrieved ordinarily, reverse engineering is argued to be a form of *corrupt hacking*; it is, therefore, a field that a few others in the electronics industry attempt to avoid. However, for many number of circuit engineers (as well as other engineers), there is brilliance in the entire concept of reverse engineering.

According to those who are not against the field, reverse engineering is a way of outsmarting an already finished product; additionally, as they would insist, isn't the point of engineering exactly that – to build and re-build until satisfying outcome is achieved? It may be considered as a form of hacking, but many contest to the idea that it is behind corrupt objectives.

XIV.D. – The Construction of Reverse Engineered Projects

Inarguably, reverse engineering is rather *destructive* and *invasive*. Some are not in favor of it since they do not welcome the idea of tearing their works apart. For others, however, it is a creative way of improving an already completed project; especially if the particular project is outstanding, they are granted the chance to make it even more outstanding.

Among the common projects that can be modified with the use of reverse engineering are *alarm clock radios, coffee machines,* and *colored lamps*. Alongside, the knowledge of internal systems, they can add a unique functionality that is not included during commercial distribution. For instance, you can add a new beeping sound to an alarm clock radio.

For the construction of various reverse engineering projects, it's advised to have a set of tools handy. Prepare a set of *screwdrivers, magnifiers,* and cutting equipment. Additionally, when the disassembly is completed, have a pack of *electrical tape* nearby.

Reverse engineering project tips for beginners:

- Take pictures of a circuit's front and back system prior to disassembly; you can use it as reference

- List down the set of procedures you plan on following; make sure you adhere to them

XIV.E. - Reverse Engineering & CAD

Reverse engineering rode along with the popularity of *CAD or Computer-Aided Design*. Through the years, the field that used to be limited to the basic improvement of a circuit's system began incorporating complex features. It provides a circuit creator the chance to analyze the internal portion of a device.

Moreover, instead of settling on getting a fundamental view of a circuit project that requires reverse engineering, a circuit creator can meticulously analyze a circuit; he can inspect each component thoroughly and come up at the most practical solution. For its development, circuit engineers, circuit designers, and other professionals on electronics started collaborative works with *architects* and those who are skilled in CAD.

Advantages of using CAD technology for a circuit project:

- To improve the alignment of each circuit component

- To enhance the designation of spaces within a circuit's system

- To modify geometric subjects for boosted performance

76

- To zoom in (and look closely) on each circuit component

XIV.F. - The Legality of the Industry

In light to the different discussions regarding its similarity to *corrupt hacking*, the reverse engineering industry is subjected to various legal complaints. There are even laws (that usually fall under contract laws and fraudulent manufacturing laws) meant for the protection of all sectors that employ *reversed* circuits or electronic devices.

The term *interoperability* is introduced in relation to a variety of legality concerns. With the emergence of cases that revolve around the disassembly of circuit's parts to compromise the quality of a circuit, some who tackle reverse engineered project are not received well.

Reverse engineering is, therefore, only considered illegal if the primary goal is to achieve interoperability. If the goal is for the improvement of a circuit's overall performance, it is encouraged; along with almost every other means of repairing a system, it is even recommended to arrive at a desired purpose.

Chapter XV – Hacking the System

Different communities of hackers host events that give light to those who are passionate in circuitry.

One community, *Artisan Asylum*, hosts a circuit hacking night once a week. In the gathering, circuit engineering fellows – from beginners to professionals, come together to discuss various concepts in circuitry and electronics. There, like-minded individuals share their love for circuitry and talk about their favorite projects, and basically, anything in the world of circuitry.

Moreover, Artisan Asylum's circuit hacking night provides great learning opportunities for circuit hacking enthusiasts. It presents lessons on how to solder, how to use particular computer programs, and how to modify a circuit to function as desired. Apart from teaching individuals the basics and advanced techniques on how to hack a circuit, and have it work as desired, the community encourages the attendees to think outside of the box and come up with brilliant and innovative ideas.

XV.A. – About Circuit Hacking

Circuit hacking, since it is linked with the word *hacking*, can sometimes be perceived as fraudulent. However, there is nothing fraudulent with it since the main reason why there are circuit hackers is for the improvement or the revision of a project; hacking, in this essence, is defined as the modification of an existing circuit to use it for a different purpose. And, in most cases, a circuit hacker is an individual who exhibits cleverness, open-mindedness, and technical aptness.

Due to a few similar concepts, circuit hacking and reverse engineering are said to be one and the same; they are not. While both may include certain techniques that are intended for the improvement of a circuit's operations, the former is merely focused on developing an existing circuit; it may be invasive, too, but it doesn't involve the *deconstruction of an entire circuit*. Especially if it was determined that the installment of a particular component can achieve a desired functionality, having to tear apart the other sections of a circuit is deemed unnecessary.

Common circuit hacking methods:

- Patching – a simple circuit hacking method that describes identifying a circuit's control mechanism or the most integral part of a circuit. Once the main component is identified, you can install a new and better component

- Component replacement – it is defined as replacing at least one component of a circuit with another component that comes with better quality

XV.B. – A Hacker's Main Tool: FIB Technology

FIB or *Focused Ion Beam* Technology is considered as one of a hacker's main tool since it grants him the chance to hack almost any circuit. Ever since the initial introduction of the applications in the 1990s, their usefulness hadn't come unappreciated. The early versions were not only quite expensive, but also, clearly, in need of improvement; later, the tools underwent continuous modifications from many electronics enthusiasts.

According to a study that was led by the engineers at Berlin Technical University, a person skilled in circuitry can install FIB Technology-based applications to hack into a system's security. For the particular research, an IC with low-level security was the focus; the objective was to work around its level of security with the goal of deliberately eliminating its defensive mechanism. The study was, of course, successful and eventually, it was proven that even high-level tools can be hacked with the same practice. And, as it follows, it sheds light on the concept that *there is no such thing as a tamper-proof circuit.*

Moreover, FIB Technology, as a clever technique of manufacturing, developing, and re-wiring a circuit, has earned the approval of different communities of hackers and circuit engineers. Alongside its advantage of boosting a system's performance, it reduces regular operation time. Due to its ingenious way of allowing an individual who's working on an electronic device to design (and even re-design repeatedly) his project, it was subjected to further developments.

Important parts of applications that incorporate FIB Technology (as shown in the layout):

- Aperture – it is in charge of gathering visual aids, then, modifying these tools for a clear display of retrievable information from a particular electronic device

- Deflector plates – it accepts, interprets, then, measures receivable data; initially, it acknowledges all information prior to the screening of the unnecessary ones

- Extractor – it is in charge of drawing out information from an electronic device, then, transferring them to the hacking mechanism

- Lens – it is in charge of making adjustments to assist when processing information

- Octupoles – it is also known as double quadrupoles or octopoles, which means something that has eight poles; it controls beams of ions

- Suppressor – it is designed to prevent power overload due to voltage spikes; it works by regulating the amount of electric current within a system so a device can remain functional

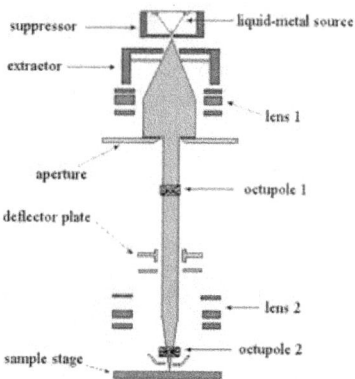

The installment of FIB Technology can be very promising to a hacker

Functions of applications that incorporate FIB Technology:

- Enable and/or disable intruder detection

- Evaluate a circuit's behavior

- Evaluate a circuit's defects

- Gather secret codes and security keys

- Obtain personal details, sensitive data, and proprietary information

- Remove protection systems (e.g. tamper networks, trace meshes, optical sensors, etc.)

- Route incoming data to be received by another network

- Trace and re-trace changes

XV.C. – Circuit Hacking Project

A plus side to the knowledge of how a circuit operates is that you can choose any from your pack of electronic devices and have it upgraded according to preference; the possibilities of its new functions are endless. You won't even have to spend a grand amount, so long as you're familiar with the functions of particular installments. The result of hacking the original system may be rather bizarre, since the product is no longer the same, but nonetheless, the result is likely the way you desired.

In the sample project below, the goal is hack a charger; particularly, modify an existing charger and have it operate with a battery. It can be connected to any electronic device with a USB port. This is useful during emergencies when an electric outlet is nowhere to be found.

Project example (derived from http://www.maximumpc.com): USB Charger with Battery

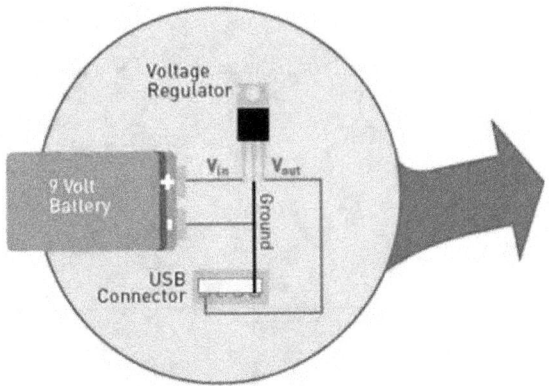

A USB charger with battery can be used for hours (depending on your voltage source)

Materials:

- Charger
- 5-volt voltage regulator
- 9-volt battery
- 9-volt battery clip
- Electrical tape

- USB connector

- Copper wire

Procedures (as shown in the layout above):

1. On a side of the charger, drill a hole for the placement of the USB connector.

2. On the other side of the charger, drill a hole for the placement of the 5-volt voltage regulator.

3. At the bottom center (between the 2 holes), place the 9-volt battery.

4. Above it, establish a ground using the copper wire and 9-volt battery clip.

5. Solder the components together.

6. Wrap the product with electrical tape.

Chapter XVI – Advanced Circuit Engineering: Microcontrollers & Robots

Among the plethora of prospects for a circuit engineer is the opportunity to engage in *circuit-bending* or the art (and science) of modifying existing circuits of electronic devices, and turning them into new musical instruments. In many cases, he isn't required to follow a set of rules for *tweaking* a circuit to incorporate a particular sound; in fact, he can re-design an electronic device as desired.

Alongside, an advantage of circuit-bending is its reward of reducing necessary expenses. In the event that he is very resourceful, the circuit creator can maximize the advantage even more; he can install used (but in working condition) components or less costly parts.

Circuit-bending is merely one of the exciting possibilities for a fellow in circuitry. The options are rather limitless, especially if you let your creativity run loose. So long as you are certain that a circuit will work given a particular arrangement, you shouldn't hold back in taking your beginner's knowledge of circuits to an advanced level.

XVI.A. – A Circuit Engineer's Future in Robotics & Computer Engineering

A reward of being skilled in circuitry is the opportunity to venture into other engineering fields such as *robotics engineering* and *computer engineering*. Your knowledge of how a circuit operates? You can look at it from a brand new perspective; you don't have to be simply in the industry of circuit engineering or electronics engineering. Apart from its offer of a more bountiful career; you can employ it to create a project (or a batch of projects) that you can be proud of. With the mastery of the basic circuitry lessons, delving into related fields becomes easier and more exciting.

With your interest in circuitry, you may seek for other career positions; usually, the employers of robotics engineers and computer engineers welcome circuit engineers into their workforce due to *trainability, familiarity with circuits,* and *good background in electronics.* It may take another set of years of studying, along with new skills to learn, but you can definitely go higher; it takes commitment from your end, too.

Advanced lessons that will be useful for a circuit engineer:

- Integral connections for hardware components

- Software and hardware essentials

- Robotic essentials

- Computer operations

- Computer architecture

- Computer programming (recommended programming languages are C and C++)

XVI.B. – Microcontroller + Microcontroller Programming

A *microcontroller* is a small device that serves as the computer in a circuit; it is a common tool that can be found in *remote controls, smart medical assistance equipment, office machines, state-of-the-art appliances,* and *engine control systems.* With the rapid pace of various information-retrieval operations, its function of addressing *size, cost, time,* and *performance* concerns offers privileges a user with improved overall performance. It is usually implanted on a separate electronic device before or after that device is finished. In certain cases, it uses 4-bit words and low frequency clock-rate operations.

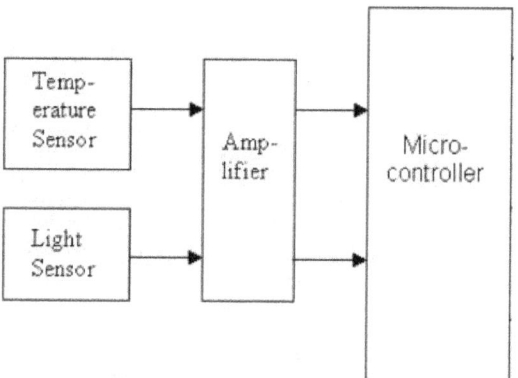

A microcontroller is dependent on a programmer's instructions

Moreover, while a microcontroller is a very useful device, it is only as good as the program that was written for it; this is where the importance of *microcontroller programming* enters the picture. It is a tool that merely executes certain instructions; in the absence of specific commands, it won't function. However, if you design it with even a hundred task capabilities, it can perform each one without fault; granted it was programmed well, it can power a device as desired.

What a microcontroller can do:

- Enable and/or disable clock feature

- Enable and/or disable light capabilities

- Configure audio and video settings

- Add an external monitor

- Automatically detect and/or repair errors

- Automatically update old components

- Magnify an electronic device's performance

- Eliminate unwanted features

- Tighten security features

Example of microcontroller programming:

Result	Microcontroller Programming (in C language)

By tweaking its USB components, a microcontroller's clock setting can be formatted. It is also programmed to activate BL or backlight with a trigger.	for (;;) [
	
	handleserial ();
	
	CDC_USBdevice (&interface)
	
	CDC_USBdevice (&digitalizer)
	
	handleserial ();
	
	]
	
	]
	
	/Setup/ [
	
	int ret
	
	USB_initialize
	
	clock_setup (div1);
	
	BL_off (1), BL_on (0);
	
	ret = digitizer_USB_initialize ();

	if (ret)

XVI.C. – Robots + Robotics

Another way that you can take your knowledge in circuitry up a notch is to consider a career in *robotics engineering*. Due to your familiarity with a circuit, you can begin honing your skills in creating coherent circuits; you can construct a series that is dependent on the individual circuits to deliver its primary function.

As you commit to robotics, like in programming, a lineup of new skills (e.g. artificial intelligence developments, dynamics, motion planning techniques, mapping tactics, etc.) must be acquired, too. This time, other than your hand in circuits and other electronic components, you have to visit aspects in the *mechanical* fields. However, given your exposure to similar concepts in circuitry, learning in the field may not be a challenge.

With your expertise in circuits, you can be behind innovative and extraordinary robotics projects that can be advantageous for both commercial and personal use. It just takes a matter of determination to move forward, and the fact remains that a course in circuit engineering can present you with a satisfying place in robotics, as well as other promising opportunities.

Conclusion

Thank you again for purchasing this book!

I hope this book was able to help you understand the fundamentals in circuit engineering. The lessons shared to you here are meant for a beginner in the subject; the different discussions are written simply. And, so far, it may have dawned on you that there's still more to discover about circuits.

The next step is to learn even more about circuitry and circuit engineering. Especially if you're considering a career in the field, advanced lessons would be good. Since this book has introduced you to the subject, and maybe inspired you to see the fun side in circuits, as well as electronics and electrical engineering, you may want to take the beginner's perspective to a whole other level.

Finally, if you enjoyed this book, please take the time to share your thoughts and post a review on Amazon. It'd be greatly appreciated!

Thank you and good luck!

Book 2
Robotics
By Kenneth Fraser

The Beginner's Guide to Robotic Building, Technology, Mechanics, and Processes

Robotics: The Beginner's Guide to Robotic Building, Technology, Mechanics, and
Processes

Table of Contents

Introduction

I want to thank you and congratulate you for purchasing the book, *"Robotics: The Beginner's Guide to Robotic Building"*.

This book contains proven steps and strategies on how to build your own robot that will perform certain functions as you want it to do.

For most people, a robot is a machine that could mimic a human such as R2D2 and C3PO in Star Wars. But these types of robots are still in the figments of our imaginations. We are still far from giving robots high level of artificial intelligence to easily adapt and interact to its environment. There is however pioneering works on artificial intelligence that hopes to create humanoid robots.

The type of robots that exist and working today are robots that are programmed to do things that are too dangerous for humans, too repetitive, or just plain messy. These robots are often found in wide range of industries and places such as oil refineries, hospitals, laboratories, factories, and even in the Outer Space. There are about more than a million robots are working in different fields today.

There are types of robots that bring joy to kids such as the popular AIBO ERS-220 that is a bestseller toy during Christmas. While some robots perform great feats by discovering new places and gathering important data in the name of science, specialized robots such as the Mars Rover Sojourner and the underwater robot Caribou are sent to places that average humans cannot go.

Robots are exciting machines to play with, but they are more exciting to build. For hobbyists, building their own robots that capable of doing whatever they program these machines to do gives them pure delight.

This book introduces you to the science of robotics – its basic elements and fundamental concepts. And at the course of your reading, you will learn all the essential aspects you need to build your own robot.

Thanks again for purchasing this book, I hope you enjoy it!

Chapter 1 – What is a Robot?

It is interesting to know that even with all the hype about robots and with all the milestones in robotics, there is still not standard definition for a robot. There are, however, some basic characteristics that a robot should have and this could help you determine if a certain object is a robot or not. It will also guide you in deciding what features you need to build into a machine before you can say that it is a robot.

Four Basic Characteristics of a Robot

A robot has four basic characteristics: sensing, movement, intelligence, and energy.

Sensing

Basically, a robot must be capable of sensing its environment much similar to the way humans sense its surroundings. Robots could either sense through light sensors that mimic the functions of the eyes, or be equipped with chemical sensors that function like the nose, sonar sensors like the nose, touch sensors like the skin, and taste sensors like the tongue. These sensors will help the robot to become aware and understand its environment.

Movement

A robot should have the ability to move around through walking on legs, rolling on wheels, or through propellers. It's either the entire robot is able to move or just some parts of it such as head, arms, or just legs.

Intelligence

A robot should be equipped with artificial intelligence or AI. This is usually done through computer programing. Hence, you need a background in programming to provide your robot with the needed intelligence. You need to program the robot's intelligence so that it will know what to sense and how to move.

Energy

A robot should have a way to power itself. The energy source could be electrical, chemical (battery), or solar. The method by which your robot energizes itself depends on what your robot is required to do.

Working Definition of a Robot

For the purpose of discussion and for reference, we define robot as a machine that contains control systems, sensors, manipulators, software and power supplies that works together to do certain tasks.

Building a robot requires understanding of the fundamental principles of mechanical engineering, mathematics, physics, and computer programming. In special cases, it also requires specific knowledge on chemistry, biology, and medicine. In studying robotics, you need to be actively engaged with wide range of disciplines to build robots that could solve certain problems.

A Brief History of Robotics

The word "robot" was first used in a play entitled R.U.R (Rossum's Universal Robots) written in 1921 by Czech writer Karl Capek. This play is about machines that are built to work on a factory and eventually revolted against their human masters. Robots are the Czech word for slave.

Meanwhile, the word robotics also first appeared in a work of fiction. Russian-born American fictionist Isaac Asimov used it in his short story "Runabout" (1942). Compared to Capek, Asimov had a more positive opinion of the role of robots in the society. In general, he described robots as useful machines that serve humans and perceived them as a "better, cleaner race. He also proposed the three Laws of Robotics:

First Law of Robotics

A robot may not injure a human or, through inaction, allow a human to come to harm.

Second Law of Robotics

A robot must obey the orders given by humans except if such orders would violate the First Law.

Third Law of Robotics

A robot may protect itself as long as such protection do not violate with the First Law or Second Law of Robotics.

Early Models of Robots

Among the earliest cases of a mechanical system designed to perform a regular task was recorded around 3000 BCE. Egyptian water clocks are added with human statuettes to hit the hour bells and signal the passing of time. In 400 BCE,

Archytus of Taremtum, who was known as the inventor of pulley and screw, created a pigeon made of wood that is capable of flying. Meanwhile, hydraulically-powered figurines that could speak prophecies were common during the Greek domination of Egypt during the second century BCE.

In the first century C.E., Petronius Arbiter built a doll that is capable of moving like a human being. In 1557, Giovanni Torriani built a wooden robot, which could fetch the Emperor's bread every day from the store. By 1700s, robotic inventions became common with numerous impractical yet ingenious machines such as steam-powered automata crafted in Canada as well as the popular talking doll by Thomas Edison. Even though these creations may have inspired the design and functions for the modern robot, the progress during the 20th century in the field of robotics exceeded previous advancements many times over.

The First Modern Robots

The robots that we are familiar with were built by George C. Devol in the 1950s. The inventor from Kentucky designed and patented a reprogrammable manipulator that he dubbed as "Unimate" derived from "Universal Automation." For years, he tried commercializing his product, but failed. But in 1960s, the entrepreneur-engineer Joseph Englberger bought the patent from Devol and modified it into an industrial robot. He established his company, Unimation, for production and marketing of these products. He was successful in this venture, and in fact, Englberger is regarded today as the Father of Robotics.

Robotics also progressed within the academic institutions. Alan Turing, pioneering computer scientist, mathematician, logician, and cryptologist, published his book "Computing Machiner and Intelligence" where he proposed a test to determine if a machine has the capacity to think for itself. This test is known as the Turing Test.

In 1958, Charles Rosen of the Stanford Research Institute created a research team to work in the development or a robot known as "Shakey" that was more advanced compared to Devol's Unimate. Shakey can move around through the room, sense light through his "eyes" move around strange environment, and to a particular degree, and react to what is happening to his surroundings. He was called Shakey because of his clattering and rickety motions.

In 1966 at Massachusetts Institute of Technology (MIT), Joseph Weizenbaum created an artificial program named ELIZA, which functions as a computer psychologist that manipulates its user's statements to formulate questions.

In 1967, Richard Greenblatt developed MacHack, a program that is capable of playing chess, as a response to a critical article written by Hubert Dreyfuss where he boasted that no computer program can beat him in chess. When the program is finished, Dreyfuss was invited to play and was defeated. This program was the foundation of future chess programs that eventually developed into Big Blue, the program that defeated Grand Master Gary Kasparov in 1997.

The interest in robotics is one of the major catalysts in the development of computers. In 1964, the IBM 360 becomes the first computer to be produced massively.

Robots are also crucial in pioneering space explorations. In 1969, the United States successfully used the latest technology in robotics and computing for Neil Armstrong's landing on the moon. Robots also helped in the expansion of scientific knowledge. In 1994, Carnegie Universities crafted Dante II, an eight-legged walking robot that successfully descends into the crater of Mt Spur to gather samples of volcanic gas.

Commercial companies also leveraged on the mass appeal of robots. In 1999, Sony released its original version of AIBO, a robotic dog that can entertain, learn, and communicate with its owner. Advanced versions have followed in the succeeding years, with the final model, the ERS-7M3, released in 2005.

Honda also released its ASIMO robot, an advanced humanoid robot in 2000. In 2004, Epsom was hailed as the world's smallest robot (7 cm high and weighs only 10 g.) The robotic helicopter is designed to fly and capture videos during natural disasters.

After being released in 2002, a robotic vacuum cleaner known as the Roomba became a huge hit. It sold more than 2.5 million units, which shows that there's really a huge demand for domestic robot technology.

Hundreds of films feature robots such as The Day the Earth Stood Still (1951), Arthur C. Clark's 2001: A Space Odyssey (1968), Star Wars (1977), Blade Runner (1982), Terminator (1984), Nemesis (1992), I, Robot (2004), Transformers (2007), and many more. The popularity of robotic films shows that people are inspired and delighted by the idea of machines that can independently move and think for itself.

If you are ready to build your own robot, continue to the next Chapter to help you get started.

Chapter 2 – Get Started

The first step in building your own robot is to determine what it should do, that is, your purpose of why you are building the robot. Robots can be used in different situations and are mainly designed to assist humans. It will help you a lot to learn first the different purposes and uses of robots.

Basically, robots are divided into two main groups: industrial and domestic robots.

Industrial Robots

Industrial robots are used in factories to manufacture products with precisions such as computers, cars, cellphones, medicine, and even food. Robots increased the productivity in different workplaces, which resulted to booming industries. Each type of industrial robot has its specific form that corresponds to its function. For instance, robotic arms are often used in car assembly lines to spray paint or weld frames. Robotic arms are among the most common robots today. Recently, agricultural robots have been introduced mainly to perform farm tasks such as cutting weeds and harvesting crops.

Domestic Robots

Domestic robots are mainly used in the home to perform household chores. They usually perform repetitive tasks every day such as vacuuming floors, mowing the lawn, vacuuming floors, and other chores that people usually don't have time to do. For example, there are vacuum robots that can clean the floors. They are equipped with motion sensors so they will not run into any object. You just need to push the switch on and it will do its job. It could pick up dust and pet hairs and could be used for hours.

There are also mower robots that could mow lawns. They are equipped with sensors to detect grass edges. Domestic robots are also used for entertainment such as Robosapien, AIBO, and iDog.

Choosing a Robotic Platform

The next step in building your own robot is to decide on the type of robot you want to build. A usual robot design usually begins with "inspiration" of what the robot will do and what it will look like.

The types of robots that you can build are endless. As long as you can envision something that a robot can do, you can work your way to achieve it. But for beginners, you can start with the following types: land robots, aerial robots, aquatic robots, stationary robots, and hybrid robots.

Land Robots

Land-based robots, particularly those added with wheels are among the most common mobile robots built by beginners, because they often require minimal investment while providing the opportunity to learn more about robotics. Meanwhile, the most advanced type of robot is the humanoid robot, which is akin to humans. Humanoids require several degrees of freedom and synchronization of different motors and use several sensors.

Wheeled Robots

Wheels are among the most common method of adding mobility to a robot and are used to mobilize many different sizes of robots and robotic platforms. Wheels could be about any size, and there's no limit in the number of wheels that you can add. More often than not, robots that are equipped with three wheels are using two wheels and a caster at one end. More advanced robots with two wheels are using gyroscopic stabilizing technology.

Meanwhile, robots that are added with four to six wheels usually use several drive motors that decreases the risk of slippage. Also, mecanum wheels or omni-directional wheels can provide the robot considerable benefits in mobility. Most beginners in robotic building are mistaken in thinking that inexpensive DC motors can mobilize robots that are medium in size. As you will learn later, there are more factors that you need to consider before you can add mobility to your robot.

Advantages

Wheeled robots are ideal for beginners as they are often more affordable to build. They have simple design and construction, and there are unlimited options. In addition, robots with six wheels or more could rival the mobility of a track system.

Disadvantages

Wheeled robots usually have small contact area, because only a small portion of the wheel is touching the ground. This results to lower traction that may cause slippage.

Tracked Robots

Tracks are used in tanks for mobility. Even though tracks, also known as treads, don't provide the added torque, they can decrease slippage and can equally distribute the robot's weight. This makes the robot easier to mobilize in loose ground such as gravel and sand. In addition, flexible track systems could easily navigate through a bumpy surface. Most hobbyists also believe that tank tracks are quite cool compared to wheels.

Advantages

Steady contact with the ground avoids slippage, which is prevalent with wheels. The track system also distributes weight evenly, which helps the robot in navigating different surfaces. Tracks can also be used to extensively enhance the ground clearance of the robot without adding a bigger drive wheel.

Disadvantages

The main disadvantage of using a track system for robots is that in turning, there's the tendency to cause damage to the surface that also causes damage to the tracks. In addition, robots are often built around the tracks, and there's a limit in the availability of the tracks. Drive sprocket can also considerably restrict the number of motors that you can use.

Legged Robots

More and more robots are using legs for movement. Legs are usually ideal to use for robots that should navigate on uneven ground. Many prototype robots are built with six legs that allow the robot for static balance. Robots with fewer legs are more difficult to balance as it requires dynamic stability. Once the robot ceases moving in the middle of the stride, it could fall over. Even though there were robots with one leg moving by hopping, bipeds, quadrupeds, and hexapods are the most common forms.

Advantages

The leg motion is the most natural among the platforms, and it can easily overcome big obstacles and move through rough surface.

Disadvantages

Most beginners are discouraged in building their first robot that moves using legs, as it requires high level of electronic, mechanical and coding skills. You also need to find a small battery that can provide the required power, so legged robots are usually expensive to build.

Aerial Robots

Humans have long been inspired by the idea of flight, and this transcends into the field of robotics. The idea of Autonomous Unmanned Aerial Vehicle (AUAV) has gained popularity over the years, and many enthusiasts have developed numerous prototypes. However, the benefits of crafting aerial robots have yet to prevail over the disadvantages. In building aerial robots, many hobbyists are still using commercial remote controllers. Professional aircrafts such as the Predator commissioned by the US military were partially autonomous though recently, updated versions of the Predator have completed aerial missions with only minimum human intervention.

Advantages

Aerial robots are great for surveillance, and remote controlled aircraft has been developed through the years, so there is a diverse community for mechanics where you can find support and know-how in building your own aerial robots.

Disadvantages

There is still limited community when it comes to autonomous control, as most of the knowledge on this field is protected by the US military. Meanwhile, this robot type is expensive as the whole robot could be broken if you miscalculate the steps and lead the robot into a crash.

Aquatic Robots

Recently, more and more hobbyists, communities, and companies are building unnamed aquatic vehicles. There are still many hindrances to overcome in order to make aquatic robots more enticing for the wider communities in robotics. But it is interesting to take note that there are companies today who are manufacturing robots that can clean pools. Aquatic robots can use thrusters, ballast, wings, tails, and fins to move under water.

Advantages

A massive part of the ocean is still unexplored so there's a lot to discover if you choose to build aquatic robots that could help in discovering the underwater world. The robot design is also guaranteed to be unique, and it could be tested in a pool.

Disadvantages

Aquatic robots are often very expensive to build, and there is the risk that the robot could be lost while deep in the ocean. You should also take note that most electronic parts don't pair well with water, especially salty water. You also need to consider the water pressure as going beyond deep sea needs considerable investment and research. There is also very limited robotic community that can provide support, and also limited wireless communication options.

Hybrid Robots

Your concept for the robot may not easily fall into any of the categories mentioned above or could be composed of various functional components. Take note that this book is written to guide you in building mobile robots and not those with fixed designs. In building a hybrid design, it is best to use a modular design where each functional component could be taken off and tested as a separate part.

Advantages

Hybrid robots are designed and built according to your preferences and needs. These robots could be used for various tasks and can be composed of modules. Hybrid robots could lead to versatility and increased functionality.

Disadvantages

Hybrid robots are often complicated to build and expensive. Parts need to be customized to fit the design.

Grippers and Arms

Even though grippers and arms don't fall under the category of mobile robots, robotics basically began with end-effectors and arms. Grippers and arms are the most ideal way for a robot to interact with the environment it is dealing with. Basic robotic arms could have just two to three motions; while more advanced arms could have more than a dozen movements.

Advantages

Most robotic arms and grippers have simple designs, and it is easy to make a three to four degree of freedom robotic arm with a turning base and two joints.

Disadvantages

Robotic arms are stationary unless you fix them on a mobile platform. The cost of building arms or grippers depends on the lifting capacity you need.

In the next chapter, you will learn how to choose the right actuators or motors for your robot.

Chapter 3 – Understanding Actuators

After learning general information about robots and robotics in the first two chapters, it is now time to choose the right actuators to mobilize your robot.

What Are Actuators?

Actuators are devices that transform energy into physical motion. In robotics, this energy is usually electrical energy. Most actuators today produce either linear or rotational motion. For example, a DC motor is a type of actuator.

Selecting the right actuator for your robot requires learning the available actuators, and some fundamental knowledge of physics and mathematics.

Rotational Actuators

Rotational actuators convert electrical energy into rotating motion. There are two primary mechanical parameters that distinguish each actuator: (a) the rotational speed that is often measured in revolutions per minute or rpm and (b) torque or the force that the devices can produce at a given distance often expressed in Oz•in or N•m.

AC Motor

Alternating Current (AC) is rarely used in robots because most of them are powered through Direct Current (DC) in form of cells or batteries. In addition, electronic parts use DC, so it is easier to use the same type of power supply for the actuators. AC motors are primary used in industrial settings where high torque is necessary or where the motors are connected to a wall outlet.

DC Motor

DC motors are often cylindrical in shape but they also come in different shapes and sizes. They also have output shafts that rotate at high speed often between 5000 and 10000 rpm. Even though DC motors rotate very fast, most have low torque. To decrease the speed and add torque, a gear could be added. To install a motor into a robot, you must fix the body of the motor to the robot's frame. Hence, motors usually have mounting holes that are basically located on the motor's face so that they can be easily installed. DC motors could either rotate in counter clockwise or clockwise. The angular movement of the turning shaft could be measured using potentiometers and encoders.

Geared DC Motor

A DC Motor could be added with a gearbox to reduce the motor's speed and enhance its torque. For instance, if a DC motor rotates at 5000 rpm and produces a 0.0005 N•m of torque, adding a 123:1 ("one hundred and twenty three to one") gear would reduce the speed by a factor of 123 (resulting to 5000 rpm / 123 = 40 rpm) and increase the torque by a factor of 123 (0.0005 x 123 = 0.0615 N•m). The most common types of gears are planetary, spur, and worm. Similar to a DC motor, a geared DC motor can also rotate in either clockwise or counter clockwise. You can add an encoder to the shaft if you want to know the number of rotations of the motor.

Hobby Servo Motors

Hobby Servo Motors, also known as R/C Servo Motors are actuators that rotate to a certain angular position, and were traditionally used in more expensive remote controlled machines for controlling or steering flight surfaces. Today, they are used in different applications so their prices have been reduced considerably, and the variety has also increased. Most servo motors can only rotate about 180 degrees. A hobby servo motor is composed of a DC motor, electronics, gears, and a potentiometer that measures the angle. The latter works with the electronics to mobilize the motor and stop the output shaft at a certain angle. In general, these servos have three wires, voltage in, control pulse, and ground. A robot servo is a recently developed servo that provides both position feedback and continuous rotation. Servos could rotate clockwise or counterclockwise.

Stepper Motors

As the name implies, stepper motors rotates following certain steps or degrees. The number of degrees the shaft rotates with every step could vary depending on various factors. Majority of stepper motors don't include gears, so similar to a DC motor, the torque is quote low. Fixing gears to a stepper motor has similar effect as installing gears to a DC motor.

Linear Actuators

Linear actuators produce linear movements. They have three primary distinctive mechanical properties: (a) the force measured in kg or lbs (b) speed measured in m/s or inch/s and (c) the maximum and minimum distance that the rod could move also known as the stroke measured in inches or mm.

Linear DC Actuator

A linear DC actuator is usually composed of a DC motor attached to a lead screw, which also turns as the motor moves. The lead screw has a traveler that is forced either away or towards the motor, basically transforming the rotating motion to a linear movement. Some DC linear actuators integrate a linear potentiometer that adds a linear position feedback.

Solenoids

Solenoids are comprised of a coil wound surrounding the mobile core. Once the coil is energized, the core is forced away from the magnetic field and creates a motion in one direction. Several coils or some mechanical arrangements will be needed to provide movements in different directions. A solenoid stroke is often very small but they are often very fast. The strength primarily depends on the size of the coil and the electrical power passing through it.

Hydraulic and Pneumatic Actuators

Hydraulic and pneumatic actuators use liquid or air respectively to create a linear movement. These actuators could have lengthy strokes, high speed and high force. To use these actuators, you need to use a fluid or air compressor that makes them harder to use compared to basic electrical actuators. These are often used in industrial applications because of their large size and high force speed.

Muscle Wire

Muscle wire is a specialized wire, which contracts when electricity passes through it. When electricity is gone and once the wire cools down, it will go back to its original length. This type of actuator is not fast, strong, or creates a long stroke. Nonetheless, it is one of the most convenient actuators to use if you need to work with smaller parts.

How to Choose the Proper Actuator for Your Robot

To guide you in choosing the actuator for certain tasks, consider answering the following questions to help you.

Take note that new innovations and technologies are always being released regularly, so nothing is permanent. Also remember that one actuator could perform various tasks in various contexts.

1. Do you need to mobilize a wheeled robot?

Drive motors should carry the weight of the whole robot and will most likely need a gear down. Majority of the robots utilize "skid steering" while trucks or cars utilize rack and pinion steering. If you prefer the skid steering, geared DC motors are recommended to use for robots with tracks or wheels. Geared motors provide constant rotation, and could have discretionary position feedback through optical encoders. Because the rotation needed is limited to a certain angle, you can choose a hobby servo motor for stirring.

2. Is there a limit on the range of motion?

If the range is restricted to 180 degrees and the needed torque is not a critical factor, a hobby servo motor is recommended. Servo motors are available in various torques and sizes and comes with angular position feedback. Majority of these motors use potentiometer, while some specialized ones use optical encoders. R/C servos are now popularly used to build small walking robots.

3. Do you need a motor to lift or turn heavy loads?

Raising a weight needs considerably more power compare to moving a weight on a flat surface. Torque should be prioritized than the speed, and it is ideal to use a gearbox with a powerful DC motor or a linear DC actuator with a high gear ratio. You can use an actuator system that could prevent the mass from falling if there is a disruption in the power source. This includes clamps or worm gears.

4. Do you need the angle to be precise?

Stepper motors that are paired with a motor controller cold provide a very precise angular motion. They are more ideal to use compared to servo motors because they provide constant rotation. But there are also high-end digital servo motors that use optical encoders and can provide high precision.

5. Do you need to achieve movements in a straight line?

Linear actuators are ideal for moving parts and placing them in a straight line. They are available in different configurations and sizes. For fast movements, you must consider solenoids or pneumatics, for high torques, you can use linear DC actuators or hydraulics, and if the movement requires minimum torque, you can use muscle wire.

Chapter 4 - Microcontrollers and Motor Controllers

Microcontrollers are considered as the "brain" of the robot because it is responsible for all decision making, computations, and communications. These are devices with the capacity to execute a program (a series of instructions).

To interact with the external world, a microcontroller has a sequence of electrical signal connections (known as pins), which could be switched on or off using programming functions. These pins are also used in reading electronic signals that are released by sensors or other devices and determine if they are low or high.

Majority of microcontrollers today could measure analogue voltage signals, or signals that could have a full range of values rather than just two specified states by using analog to digital converter or ADC. Through the use of ADC, a microcontroller could assign a numerical value to the analog voltage that is neither low nor high.

What Could Microcontrollers Do?

Numerous complicated actions could be achieved by setting the pins low and high creatively. Nonetheless, building complicated algorithms such as smart movements and data processing or complicated programs are not yet on the range of microcontrollers because of its natural speed and resource limitations.

For example, to light a blinker, you can program a repeating sequence in which the microcontrollers could turn a pin high, wait for several seconds, turn it low, wait for several seconds and goes back to the first sequence. A light that is connected to the pin will then blink open-endedly.

Similarly, microcontrollers could be used to take control of other electronic devices including actuators when they are installed to motor controllers, Bluetooth or WiFi interfaces, storage devices, and many more. Because of its versatility, microcontrollers could be found in common everyday products. Basically, every home electrical device or home appliance utilizes at least one microcontroller.

Not similar to microprocessors found in Central Processing Units in personal computers, microcontrollers don't need peripherals such as external storage devices or external RAM to operate. Hence, even if the microcontrollers are less powerful compared to microprocessors, building circuits and products based on microcontrollers is an easier task and a lot more affordable, because minimal hardware parts are needed. Remember, microcontrollers can output minimum amount of electrical power through pins. Hence, a generic microcontroller cannot power solenoids, power electrical motors, large lights, or other direct loads. Doing this could cause physical damage to the controller.

Programming Microcontrollers

There's no need to shy away from programming microcontrollers. Not similar in the past where making a blinker took comprehensive knowledge of microcontroller and at least a dozen line of code, programming microcontrollers is fairly easy today. You can use the simplified Integrated Development Environments (IDE), which uses modern languages, full line archives that could cover all of the most common actions, and several handy samples to help you get started. You can learn more about programming your robot in Chapter 6.

How to Choose the Proper Microcontroller for Your Robot

You will need a microcontroller for any robotic building project unless you're into BEAM robotics or you want to control your robot through an R/C system or a tether. For starters, selecting the right microcontroller could seem like a difficult job, particularly considering the product range, specifications, and applications. There are various microcontrollers available today such as BasicATOM, POB Technology, Pololu, Arduino, BasicX, and Parallax.

The following questions could guide you in choosing the right microcontroller:

1. Which microcontroller is widely used for your type of robotic project?

Building robots is not a popularity contest, but the fact that a microcontroller has a large supporting community or has been used in the same project can make the design phase easily. With this, you can benefit from other experience and hobbyists. It is common for hobbyists to share codes, pictures, instructions, and videos even lessons learned.

2. Do you need specific accessories for a certain microcontroller?

If your robot has special needs or there is a certain accessory or component that is important for your design, selecting a compatible microcontroller is clearly essential. Even though most accessories and sensors could be directly interfaced with most microcontrollers, some accessories are designed to interface with a particular microcontroller.

3. Do you need special features for your robot?

A microcontroller should be able to perform all the special actions needed for your robot to function well. Some features are common to all microcontrollers such as being able to execute basic mathematical operations, having digital inputs and outputs, and making decisions. Others may need certain hardware such as PWM, ADC, and communication protocol support. You must also consider pin counts, memory and speed requirements

Motor Controllers

Motor controllers are electronic devices that serve an intermediary device between a microcontroller, the motors, and the power supply.

Even though the microcontroller decides the direction and the speed of the motors, it doesn't have enough power to directly drive them. Meanwhile, the motor controller can supply the current at the needed voltage but doesn't have the capacity to decide how fast the motor must turn.

Hence, the microcontroller and the motor controller must work together to make the motors move accordingly. The microcontroller can provide instructions to the motor controller on how to power up the motors through a standard and basic communication method such as PWM and UART. In addition, some motor controllers could be manually regulated using an analog voltage often created through a potentiometer.

The size and weight of a motor controller may greatly vary from a device that is smaller than the tip of a pencil to a huge controller that could weight several kilos. The size and weight often has a minimum effect on the robot, unless you want to build unnamed aerial or aquatic robots.

Types of Motor Controllers

Because there are several types of actuators (as we have discussed in Chapter 3), there are also several types of motor controllers: brushed DC motor controllers, brushless DC motor controllers, servo motor controllers, and stepper motor controllers.

How to Choose a Motor Controller

You can only choose a motor controller after you have decided on what type of actuator you want to use. In addition, the current that a motor draws depends on the torque it could provide. A small DC motor will not use much current, but cannot also release much torque, while a bigger motor could release higher torque but will need increased current.

Chapter 5 - Controlling Your Robot and Use of Sensors

Based on our definition of a robot, it should gather data about its surroundings, make smart decisions and then execute actions based on calculations. This also includes the option for the robot to become semi-autonomous (with aspect that are controlled by humans and other aspects that it can do on its own).

One good example of this is a complex aquatic robot. A human controls the basic motions of the robot while an installed processor measures and reacts to the underwater currents to keep the robot in one position while still preventing a drift. A camera installed in the robot would send videos back to the human while the sensors could track the water pressure, temperature, and more. Once the communication line falters between the robot and the human, an autonomous program could take over to instruct the robot to reach for the surface.

In controlling your robot, you need to figure out the level of autonomy. You need to choose if you want the robot to be tethered, wireless, or autonomous.

Tethered

Direct Wired Control

The simplest way to control a vehicle is by using a handheld controller that is physically connected to a vehicle using a tether or a cable. Knobs, switches, joysticks, buttons, and levers on the controller will allow you to control the robot without the need to add sophisticated electronics. In this setting, the power source and motors could be directly connected with a switch to control the rotation. These machines often have no artificial intelligence and are regarded as remote controlled devices than robots.

Wired Computer Control

Another method is to integrate a microcontroller into the machine but still using a tether. Attaching the microcontroller to your computer's ports will allow you to control the actions using the keyboard, a joystick, a keypad, or other device. Adding a microcontroller to your robot project may also require programming how the robot will respond to the input.

Ethernet

Another way to use computer control is to use an Ethernet interface. A robot that is directly connected to a router can also be used for mobile robots. Building a robot, which can communicate through the internet could be sophisticated, and usually a wireless internet connection is more recommended.

Wireless

Infrared

You can ditch away cables and wires if you use infrared transmitters and receivers. This is often a great achievement for beginners. Infrared control needs "line of sight" to function. The receiver should have the ability to see the transmitter to receive the data. Infrared remote controls can be used to send commands to infrared receivers that are paired with microcontrollers that interpret these signals and control the actions of the robot.

Radio Frequency

Remote control units often use microcontrollers in the receiver and transmitter for data transmission through radio frequency. The receiver box usually has a printed circuit board (PCB) that includes a small servo motor controller and a receiving unit. RF communication needs a transmitter matched with a receiver or a transceiver. RF doesn't need clear line of sight and could also provide considerable distance. Basic RF devices could allow for data transfer between devices between long distances, and there's no limit to the range for more RF devices.

Bluetooth

Bluetooth is a type of Radio Frequency and follows certain protocols for sending and receiving data. Standard Bluetooth range is usually restricted to about 10 meters although it has the advantage of controlling the robot though Bluetooth-enabled devices including laptops, smartphones, and PDAs. Similar to RF, Bluetooth provides two-way communication.

WiFi

Recent development in wireless technology enables you to control a robot through the Internet. To build a WiFi robot, you must have a wireless router that is connected to the internet and a WiFi device on the robot. You can also use a device, which is enabled with TCP/IP with a wireless router.

Autonomous

High-level robots are autonomous. With recent developments, you can now use the microcontroller in its full potential and program it to respond to input from the sensors. Autonomous control may come in different types: restricted sensor feedback, pre-programmed with no feedback from the environment, and complex sensor feedback. Genuine "autonomous" control includes different sensors and code to allow the robot to figure out by itself the smartest action to be taken in any situation.

The most sophisticated methods of control presently used on autonomous robots are auditory and visual commands. For auditory control, a robot will react to the sound of the human's voice for instructions such as "get the ball" or "turn left." For visual command, a robot may look to an object to decide on what to do. Instructing a robot to turn to the right by showing a drawing of an arrow that is pointing to the right requires complicated programming. Even though these things are no longer impossible, they need a sophisticated level programming and usually hundreds of hours.

Not similar to humans, robots are not restricted to just sound, sight, smell, touch, and taste. Robots use different electromechanical sensors to understand and discover their surroundings. Mimicking a natural organism's senses is presently a great challenge, so developers and robotic builders are using alternatives to these natural senses.

Chapter 6 - Assembling and Programming a Robot

After learning all about the fundamental blocks in building a robot, the next stage is the designing and building of the frame that will keep all the components together and will provide your robot a definite look and shape.

Constructing the Frame

There's no fix method in creating a frame, because there is often a trade-off to be constructed. You may prefer to use a lightweight frame but you may need to use costly materials. You may like a strong or big chassis but you may realize it is expensive, hard, or heavy to produce. The frame could be complicated and may take some time to design and build.

Materials

There are different materials that you can use in creating a frame for your robot. As you try different materials to construct not only robots but other types of machines, you will also understand the advantages and the disadvantages as to which material is the most suitable for a specific project. The roster of suggested building materials below comprises only the more common one, and when you have tried several of these materials, you can start experimenting or blending some together.

Basic Construction Materials

Some of the most basic construction materials could be used to build good-quality frames. The cheapest materials is the cardboard that you can usually find for free and could easily be bent, cut, layered, or bent. You can construct a reinforced cardboard box that looks a lot nicer and is more proportional when it comes to the size of your robot. You can then paint it with glue or epoxy to make it stronger then add extra layer of paint.

Structural Flat Materials

For a more durable frame, you can use a standard structural material such as a sheet of plastic, metal, or wood. You just need to puncture some holes to connect the electronic components. A stronger piece of wood has a tendency to be heavy and thick, while a thin sheet of metal could be too flexible. You can attach components to both sides and the wood will still remain solid and intact.

If you're at the stage where you are ready to have an outsourced frame, the best option is to acquire the part precision cut through a water or laser jet. Hiring a third-party to produce a custom part is recommended only if you are completely sure of the dimensions, because the mistakes could be expensive. Companies that offer computer controlled cutting services may also provide different other services such as painting and bending.

3D Printing

Building a frame constructed from 3D printed panels is not always the most structurally sound option, primarily because it is built up in several layers. However, this process could produce complex and detailed shapes that could be impossible to build using other methods. One 3D printed component may contain all the important mounting points for all mechanical or electrical parts without compromising the robot's weight. In the past decade, the cost of 3D printing is quite expensive, but as it becomes popular, the price of producing the components is also expected to go down.

Assembling the Parts Together

With the available options for materials and methods, you can now start assembling the parts together. You can follow the steps below to build a simple, aesthetic, and structurally reliable robot frame.

1. Decide on the material you want to use.

2. Gather all the parts that your robot will need, both mechanical and electrical and measure them. In case you don't have all the components ready, you can refer to the dimensions that are often supplied by the manufacturers.

3. Think of and draw various designs for your frame. It's fine not to provide details.

4. Once you find the suitable design, be certain that the structure is reliable and all the parts would be supported in the frame.

5. Sketch every component of the robot on cardboard or paper at true scale. You can also draw the parts in the CAD software and print them.

6. Test the design in CAD and in actual setting using your paper prototype by test fitting every component and connection.

7. Measure the dimensions again and when you are completely certain that your design is right, begin cutting the frame into the material. Take note to measure two times and only cut once.

8. Test fit every part before assembling the frame if in case you need some changes.

9. Construct the frame using appropriate assembling materials such as glue, nails, screw, duct tape, or any appropriate binding tools that you prefer.

10. Fit all the parts into the frame and there you go, you have just built your robot!

Constructing the Robot Parts

The last step discussed above should be described further. In the past chapters, you have already chosen the electrical parts including the actuators, microcontroller, and motor controller. The next step is to construct them so they will work together.

In the following section, we'll use standard cable colors and terminal names that encompass common parts. You must rely on manuals and datasheets when you are working on your specific parts.

Attaching Motor Controllers to Motors

A geared DC motor or a linear DC actuator usually has two wires: black and red. Attach the black wire to the M- terminal on the DC motor controller and the red wire to the M+ terminal. Connecting the wires the other way around will only cause the motor to rotate in the opposite direction. Meanwhile, servo motors have three wires: red, black, and yellow. A servo motor controller comes with pins that are matching these wires so you can just plug it directly.

Attaching Microcontroller to Motor Controllers

Microcontrollers can communicate with motor controller in different ways: 12C, R/C, Serial, or PWM. Be sure to refer to the manual for each microcontroller for specific instructions on proper connection. Regardless of the method you choose, the microcontroller and the logic of the motor controller should share matching ground reference. This can be achieved by attaching the GND pins together. Meanwhile, a logic level shifter is needed if the devices don't share the same logic levels.

Attaching Batteries to a Microcontroller or a Motor Controller

Majority of the motor controllers available today have two screw terminals for the battery labels marked with B- and B+. If the batteries you got are provided with a connector and the controller comes with screw terminals, you could still search for a pairing connector with wires that you can attach to the screw terminal. If this is possible, you need to find another way to link the battery to the motor controller while you can still unplug the battery and link it to a charger. It's possible that not all the electrical and mechanical components you have selected for your robot could operate on a single voltage, and so may need several voltage regulation circuits or batteries.

If you are building a robot with a microcontroller, DC gear motors, and maybe servo motors, it's easy to see how a battery may not be able to power every component directly. Nevertheless, it is best to choose a battery that can directly power as many devices that you need. The battery with the largest capacity must be connected with the drive motors. For instance, if the motors you select are rated a nominal 12 volts, the primary battery must also be 12 volts. So you can use a regulator to energize a 5 volts microcontroller. LiPo and NiMH batteries are the top choices for small to medium robots. Select NiMH if you need cheaper batteries and LiPo if you need light weight batteries. Always take note that batteries are powerful devices that could easily burn your circuits if they are not properly connected. Always make sure that the polarity is correct and that your device could handle all the energy supplied by the battery. If you are not certain, never make assumptions.

Adding Electrical Parts to Frame

You can attach electrical components to your frame through different methods. Make certain that whatever method you use, don't conduct electricity. Usual methods include screws, hex spacers, Velcro, double-sided tape, cable ties, glue, and many more.

Programming Your Robot

Programming is often the last step in building your robot. If you have followed the steps described in the previous chapters, by now you have selected the electrical components such as actuators, microcontrollers, motor controllers, sensors, and more. At this point, you might have already constructed your robots and hopefully it looks something like you want it to be. But without the proper program, your robot is just a cool paperweight.

It requires another book to teach you robotic programming. Instead, this section will guide you on how to get started and what you should learn.

There are several programming languages that you can use to program the microcontroller that will serve as the brain of your robot. The following are the most common programming languages you can choose:

Assembly

This programming language is just a shy away from programming a full-pledged computer, and so it could be difficult to use. This language is ideal to use if you really need to ensure complete instruction-level control of your robot.

Basic

Basic is one of the most common programming languages for robot hobbyists. This is often used in programming microcontrollers mainly for educational robots.

C++

C++ is a very popular programming language. It provides top=level functionality while you are keeping a good low-level control. A variant of C++ is Processing, which includes simplified codes to make the programming easier.

Java

Java is more developed compared to C++ and offers any safety features to the disadvantage of low-level control. Some producers of microcontrollers such as Parallax are making components for specific use with Java.

Python

Python is one of the most popular languages for scripting. It is easy to learn and could be used to quickly and efficiently integrate several programs.

If you have selected a hobbyist type of microcontroller from a known producer, there's a chance that you can find a book that you can read so you can learn how to program in their preferred programming language. But if you instead prefer a microcontroller from a smaller producer, it is crucial to see what language the controller wants to use and what tools are available.

Conclusion

Thank you again for purchasing this book!

I hope this book was able to help you to learn the basic building blocks of robot building.

The next step is to expand your knowledge in robotics, especially learning advanced programming for your robot.

Finally, if you enjoyed this book, please take the time to share your thoughts and post a review on Amazon. It'd be greatly appreciated!

Thank you and good luck!

www.ingramcontent.com/pod-product-compliance
Lightning Source LLC
Chambersburg PA
CBHW070718210526
45170CB00021B/588